Digital Literacies in Foreign and Second Language Education

Digital Literacies in Foreign and Second Language Education

Edited by

JANEL PETTES GUIKEMA
Grand Valley State University

LAWRENCE WILLIAMS
University of North Texas

CALICO Monograph Series, Volume 12

CALICO Monograph Series

Volume 2, 1991
Interactive Videodisc: The "Why" and the "How"
Edited by Michael Bush, Alice Slaton, Miguel Verano, and Martha Slayden

Volume 3, 1995
Thirty Years of Computer Assisted Language Instruction
Edited by Ruth H. Sanders

Volume 4, 1997
Nexus—The Convergence of Language Teaching and Research Using Technology
Edited by Kathryn Murphy-Judy

Volume 5, 2006
Calling on CALL: From Theory and Research to New Directions in Foreign Language Teaching
Edited by Lara Ducate and Nike Arnold

Volume 6, 2007
Preparing and Developing Technology-proficient L2 Teachers
Edited by Margaret Ann Kassen, Roberta Z. Lavine, Kathryn Murphy-Judy, and Martine Peters

Volume 7, 2008
Opening Doors through Distance Language Education: Principles, Perspectives, and Practices
Edited by Senta Goertler and Paula Winke

Volume 8, 2009
The Next Generation: Social Networking and Online Collaboration in Foreign Language Learning
Edited by Lara Lomicka and Gillian Lord

Volume 9, 2010
CALL in Limited Technology Contexts
Edited by Joy Egbert

Volume 5 (Second Edition), 2011
Present and Future Promises of CALL: From Theory and Research to New Directions in Language Teaching
Edited by Lara Ducate and Nike Arnold

Volume 10, 2012
Technology Across Writing Contexts and Tasks
Edited by Greg Kessler, Ana Oskoz, and Idoia Elola

Volume 11, 2013
Design-Based Research in CALL
Edited by Julio C. Rodríguez and Christina Pardo-Ballester

Published by the Computer Assisted Language Instruction Consortium (CALICO)
Texas State University, 601 University Drive, San Marcos, Texas 78666 USA (calico.org)

© CALICO 2014
Series first published in 1986
Printed in the United States of America

ISBN: 978-0-9891208-9-0

ALL RIGHTS RESERVED. No part of this work may be reproduced or used in any form or by any means—graphic, electronic, or mechanical, including photocopying, recording, taping, web distribution or information storage and retrieval systems—without the written permission of the publisher.

Contents

Janel Pettes Guikema Lawrence Williams	**Introduction** Digital Literacies from Multiple Perspectives	1
Heather Lotherington Natalia Ronda	**Chapter 1** 2B or Not 2B? From Pencil to Multimodal Programming: New Frontiers in Communicative Competencies	9
Lawrence Williams Lee B. Abraham Evan D. Bostelmann	**Chapter 2** A Survey-Driven Study of the Use of Digital Tools for Language Learning and Teaching	17
Silvia Benini Liam Murray	**Chapter 3** Challenging Prensky's Characterization of Digital Natives and Digital Immigrants in a Real-World Classroom Setting	69
Juan Pablo Jiménez-Caicedo María Eugenia Lozano Ricardo L. Gómez	**Chapter 4** Agency and Web 2.0 in Language Learning: A Systematic Analysis of Elementary Spanish Learners' Attitudes, Beliefs, and Motivations about the Use of Blogs for the Development of L2 Literacy and Language Ability	87
Malgorzata Kurek Mirjam Hauck	**Chapter 5** Closing the Digital Divide: A Framework for Multiliteracy Training	119
Lilian Mina	**Chapter 6** Enacting Identity through Multimodal Narratives: A Study of Multilingual Students	141
Jonathon Reinhardt Chantelle Warner Kristin Lange	**Chapter 7** Digital Games as Practices and Texts: New Literacies and Genres in an L2 German Classroom	159

Ana Oskoz Idoia Elola	**Chapter 8** Integrating Digital Stories in the Writing Class: Toward a 21st-Century Literacy	179
Carl S. Blyth	**Chapter 9** Exploring the Affordances of Digital Social Reading for L2 Literacy: The Case of eComma	201
Maarit Mutta Sanna Pelttari Leena Salmi Aline Chevalier Marjut Johansson	**Chapter 10** Digital Literacy in Academic Language Learning Contexts: Developing Information-Seeking Competence	227
Martine Peters Mary Frankoff	**Chapter 11** New Literacy Practices and Plagiarism: A Study of Strategies for Digital Scrapbooking	245
Janel Pettes Guikema Mandy R. Menke	**Chapter 12** Preparing Future Foreign Language Teachers: The Role of Digital Literacies	265
	Index	287

List of Contributors

Lee B. Abraham
Lee B. Abraham is Lecturer in Language (Spanish) in the Department of Latin American and Iberian Cultures at Columbia University (New York, USA). His research focuses on the integration of new technologies in language teacher education, language variation in new media, and multilingualism in the linguistic landscapes of the Americas and Europe.
lba2133@columbia.edu

Silvia Benini
Silvia Benini is a Ph.D. researcher and an Italian language teacher in the School of Languages, Literature, Culture and Communication at the University of Limerick (Ireland). She has a *Laurea in Lettere Moderne* (University of Florence) and the postgraduate diploma DITALS 2 in Teaching Italian as Foreign Language (University of Siena). She has been involved in various research projects on language learning strategies and theatre and Italian teaching pedagogy. Her main research interests are SLA, CALL, blended learning, language policy and planning, language revitalization.
Silvia.Benini@ul.ie

Carl S. Blyth
Carl S. Blyth (Ph.D., Cornell University) is Associate Professor of French Linguistics and Director of the Center of Open Educational Resources and Language Learning (COERLL) at the University of Texas at Austin (USA). His research interests include computer-mediated discourse, corpus linguistics, cross-cultural and intercultural pragmatics, and pedagogical grammar. He has published on metalinguistic awareness, native and nonnative role models for language learning, L2 narrative discourse, pedagogical norms, stance taking in interaction and open models for educational publishing.
cblyth@mail.utexas.edu

Evan D. Bostelmann
Evan D. Bostelmann is currently in the Master of Arts in Language, Literature, and Translation program at the University of Wisconsin at Milwaukee (USA), where he teaches French. His research interests include sociolinguistics, second language acquisition, and technology-infused language learning.
bostelm3@uwm.edu

Aline Chevalier
Aline Chevalier is Professor of Cognitive Psychology and Ergonomics in the CLLE Lab of the University of Toulouse (France). Her research focuses on the role of cognitive processes involved in the use of internet and the cognitive activity of designers of information systems.
aline.chevalier@univ-tlse2.fr

Idoia Elola
Idoia Elola is Associate Professor of Spanish and Applied Linguistics & Second Language Studies at Texas Tech University (USA). Her research focuses mainly on second language writing, such as collaborative and individual writing when using Web 2.0 tools, Spanish heritage language learners' writing processes, and revision and feedback.
idoia.elola@ttu.edu.

Mary Frankoff
Mary Frankoff is a tenured Sociology professor at Cegep Heritage College (Canada) in the Social Sciences Department. Her research interests include the "digital divide" among postsecondary students as it pertains to information-seeking activities and, consequently, academic success.
mfrankoff@cegep-heritage.qc.ca

Ricardo L. Gómez
Ricardo L. Gómez is Assistant Professor of Quantitative Research and Evaluation Methods in the College of Education at Universidad de Antioquia in Medellín, Colombia. He holds a doctorate in Education Policy and Leadership from the University of Massachusetts, Amherst (USA). His research interests include program monitoring and evaluation, empirical analysis of higher education policies, participation of underrepresented groups in academia and advanced research careers in science and engineering (STEM), and impact of national assessment systems on teaching and learning.
ric.leon.gomez@gmail.com

Janel Pettes Guikema
Janel Pettes Guikema is Associate Professor of French in the Department of Modern Languages and Literatures at Grand Valley State University (Michigan, USA). Her research interests include foreign language teacher education, language learning and technology, and the development of multiliteracies in foreign language learning.
pettesj@gvsu.edu

Mirjam Hauck
Mirjam Hauck is Senior Lecturer and Associate Head of the Department of Languages (Faculty of Education and Language Studies) at the Open University (UK). She has written numerous articles and book chapters on the use of technologies for the learning and teaching of languages and cultures covering aspects such as task design, tutor role and training, and digital literacy skills. From 2009 to 2012, she served on the CALICO executive board and is a member of the EUROCALL executive committee.
Mirjam.Hauck@open.ac.uk

Juan Pablo Jiménez-Caicedo
Juan Pablo Jiménez-Caicedo is a lecturer in the Department of Latin American and Iberian Cultures at Columbia University (New York, USA). His research draws on a sociocultural framework, Systemic Functional Linguistics and mixed method approaches focusing on language teaching and learning processes, and on the uses of multimodal and Web 2.0 technologies for the development of learners' second language academic literacies.
jj2415@columbia.edu

Marjut Johansson
Marjut Johansson is Professor of French Language at the University of Turku (Finland). Her research interests lie in the areas of sociopragmatics, dialogical linguistics, discourse analysis, and sociolinguistics. Her main area of research is media discourse and interaction. She is interested in multilingualism and language policies, as well as foreign language teaching and learning in higher education.
marjut.johansson@utu.fi

María Eugenia Lozano
María Eugenia Lozano is Associate in Spanish and Latin American Cultures at Barnard College. She is a doctoral candidate in the Language, Literacy and Culture program in the College of Education at University of Massachusetts, Amherst. Her research interests include second language acquisition, language maintenance among immigrants, and the use of Web 2.0 technologies for language teaching.
mlozano@barnard.edu

Malgorzata Kurek
Malgorzata Kurek is Assistant Professor in the Department of Languages at Jan Dlugosz University (Czestochowa, Poland). Her research interests include telecollaboration, task design in CALL, foreign language multiliteracy and CALL teacher training. She is an author of several ICT-enhanced courses for teachers and teacher trainees, and numerous presentations and publications in the field of Computer-Assisted Language Learning. She holds an MA in TEFL and a Ph.D. in CALL.
gkurka@gmail.com

Kristin Lange
Kristin Lange is a Ph.D. student in Second Language Acquisition and Teaching at the University of Arizona where she teaches courses in the Department of German Studies. Her research interests focus on L2 pedagogy, including the use of authentic L2 material, intercultural competence, and vocabulary teaching and learning.
klange@email.arizona.edu

Heather Lotherington
Heather Lotherington is Professor of Multilingual Education at York University in Toronto (Canada), where she teaches in the Faculty of Education and in the Graduate Program in Linguistics and Applied Linguistics. She is now working with Natalia Ronda and an international team to systematically investigate how communicative competence can be revised for the digital age. Professor Lotherington's most recent book is: *Pedagogy of multiliteracies: Rewriting Goldilocks* (Routledge, 2011).
hlotherington@edu.yorku.ca

Mandy R. Menke
Mandy R. Menke is Assistant Professor of Spanish in the Department of Modern Languages and Literatures at Grand Valley State University (Michigan, USA). She teaches a variety of courses, including Spanish language, linguistics, and foreign language methodology. Her research interests include language immersion education, second language phonology, and foreign language pedagogy.
menkem@gvsu.edu

Lilian Mina
Lilian Mina (Ph.D. in Composition & TESOL, Indiana University of Pennsylvania, USA) has published and presented on corpus linguistics in EFL contexts, international students' identity, peer feedback in L2 writing, community-based projects in engineering classes, inquiry-based research in the composition class, and technology in teaching writing. Her research interests include writing pedagogy, new media technologies, multimodality, virtual identity, international students, and professionalization of graduate students.
Lilian.mina@gmail.com

Liam Murray
Liam Murray teaches courses in the School of Languages, Literature, Culture and Communication at the University of Limerick (Ireland) on CALL, digital games-based language learning, cyberculture, e-learning and evaluation at both undergraduate and postgraduate levels. Areas of research interest include CALL, DGBLL and the application of social media and blog writing to second language acquisition. He is at present software reviews editor of the international journal *ReCALL*.
Liam.Murray@ul.ie

Maarit Mutta
Maarit Mutta is Adjunct Professor (Ph.D.) in the Department of French Studies at the University of Turku (Finland). She is a specialist in second language learning and teaching and is involved in educational issues as a teacher educator at the university level. Her research focuses on the problems of the online management of knowledge related to multilingual learners and their metalinguistic awareness.
maamut@utu.fi

Ana Oskoz
Ana Oskoz is Associate Professor of Spanish at the University of Maryland, Baltimore County (USA). Her research focuses on language and technology, such as the use of synchronous and asynchronous communication tools for second language learning to enhance second language writing and foster intercultural competence development.
aoskoz@umbc.edu

Sanna Pelttari
Sanna Pelttari is Lecturer in the Department of Spanish Studies, University of Turku (Finland). Her research interest has been mainly in Applied Linguistics, specifically in the use of contrastive perspective in language teaching and learning and in the improvement of language skills with the help of a particular treatment of errors.
sanna.pelttari@utu.fi

Martine Peters
Martine Peters is Professor of Education at the Université du Québec en Outaouais (Canada), where she is the Ph.D. Program Director. Her research focuses on technopedagogy and the use of technology in the writing process.
martine.peters@uqo.ca

Jonathon Reinhardt
Jonathon Reinhardt is Assistant Professor of English Language/Linguistics at the University of Arizona (USA) and Co-Director of the Games to Teach Project at the Center for Educational Resources in Culture, Language, and Literacy. His research focuses on the application of literacies-informed frameworks to L2 teaching and learning, especially in digital game and social media contexts. His co-authored book *Language at Play: Digital Games in Second and Foreign Language Teaching and Learning* appeared with Pearson in 2013.
jonrein@email.arizona.edu

Natalia Ronda
Natalia Ronda currently holds a research position at the Council of Ministers of Education (Canada). Her research examines digital literacies, broadly defined, particularly the use of new digital spaces for literacy development. Her work cross-cuts language and literacy, social media, and digital pedagogy, and specifically develops the connections between informal practices of digital media use and formal curriculum. She received her Ph.D. in education from York University in Toronto (Canada).
natalia@ronda.ca

Leena Salmi
Leena Salmi is Lecturer of French Translation at the School of Languages and Translation Studies, University of Turku (Finland). Her research interests focus on different aspects of the translator's work: search for background information as a part of the process, postedition of machine-translated texts, and the amount of translated text in everyday life. She is currently the vice-president of the Authorized Translator's Board.
leena.salmi@utu.fi

Chantelle Warner
Chantelle Warner is Assistant Professor of German and Second Language Acquisition and Teaching at the University of Arizona (USA) and co-director of the Center for Educational Resources in Culture, Language, and Literacy. Her research interests lie in foreign language literacy, with a focus on language play and aesthetics, stylistics, and literary pragmatics. Her book *The Pragmatics of Literary Testimony: Authenticity Effects in German Social Autobiographies* appeared with Routledge in 2013.
warnerc@email.arizona.edu

Lawrence Williams
Lawrence Williams is Associate Professor of French and Applied Linguistics at the University of North Texas (USA), where he is Associate Chair of the Department of World Languages, Literatures, and Cultures. His research focuses on sociolinguistics—primarily French—and the use of new technologies for communication and language teaching/learning.
lawrence.williams@unt.edu

Introduction

Digital Literacies from Multiple Perspectives

JANEL PETTES GUIKEMA
Grand Valley State University (USA)

LAWRENCE WILLIAMS
University of North Texas (USA)

1. The Role of Digital Literacies in Foreign and Second Language Education

This volume on digital literacies aims to provide CALL/SLA scholars and foreign language teachers with a focused set of studies that reflect current thinking on digital literacies from multiple perspectives and to offer recommendations for best practices in the classroom. As Kern (2000) points out, "it is precisely because literacy is variable and intimately tied to the sociocultural practices of language use in a given society that it is of central importance in our teaching of language and culture" (p. 25). Moreover, Swaffar and Arens (2005) assert that changes over the past few decades "reflect an understanding of literacy as socially bounded and contextual, no longer accessible solely through command of language as traditionally presented in many [foreign language] classrooms" (p. 3). Given the many changes in thinking and practice that have evolved over the past two decades, it now seems vital to (re)examine policies, practices, theories, and beliefs about digital literacies in foreign language education.

> These changes affect the ways we use language as well as the ways we learn languages. They also challenge our traditional understanding of literacy, which goes well beyond the skills of encoding and decoding texts. The challenges of multiculturalism and multimodal forms of communication call for a revised definition of literacy that goes beyond textual paraphrase as an adequate measure of reading ability and error-free prose as a measure of writing skills. Literacy redefined must encompass complex interactions among language, cognition, society, and culture. (Kern & Schultz, 2005, p. 382.)

Although literacy seems like a basic component of most educational endeavors, it does not necessarily enjoy a prominent place in foreign language learning and teaching. Even when literacy receives some curricular emphasis, it is often viewed as a set of basic skills for beginning learners to master, involving not much more than being able to read, write, and "sound out" all the letters or characters of a foreign alphabet or writing system (Pennycook, 2001). Likewise, digital literacy in the K-12 setting is often viewed as a set of skills that will be learned over time as a student progresses through a state-designed curriculum that includes checklists as the primary or sole evaluation of literacy-based competence and performance. In higher education, the foreign language curriculum is usually determined by textbooks, some of which seem to ignore the great potential for integrating literacy into learning modules and projects.

Unfortunately, the multimodal, visual, written, and audio texts provided in some textbook programs are often little more than tools that teachers can use for comprehension checking. While it is certainly important (and central to language acquisition) for learners to understand what they read, view, and/or hear, a review of most textbook programs reveals that a linguistic perspective of literacy dominates, resulting in a scope of literacy-based activities that is essentially limited to phonetics, morphology, syntax, and semantics. Notably missing are the cultural, pragmatic, and sociolinguistic dimensions of language and communication as well as other important notions such as, for example, learner agency (see Belz, 2002; Mercer, 2011).

Nonetheless, Kern's (2000) model for integrating literacy into the foreign language curriculum involves three perspectives: linguistic, cognitive, and sociocultural, all of which are interdependent and deserve equal attention. While there are certainly other models for promoting digital literacy in foreign language education (e.g., Byrnes & Kord, 2001; Kramsch & Nolden, 1994; Swaffar & Arens, 2005), the model developed by Kern is just one example of a call for an expanded view of literacy (including digital literacies), which is what this volume offers to CALL/SLA scholars and teachers. As Buckingham (2008) argues:

> rather than simply adding media or digital literacy to the curriculum menu or hiving off information and communication technology into a separate school subject, we need a much broader reconceptualization of what we mean by literacy in a world that is increasingly dominated by electronic media. (p. 88)

This broader reconceptualization of literacy inherently includes the digital, exemplified in constructs such as new literacies (e.g. Gee, 1999; Lankshear & Knobel, 2011) and multiliteracies (The New London Group, 1996), which focus on literacy as social practice and address the complexity of meaning-making in a world where communication is increasingly multimodal. According to Knobel and Lankshear (2008), digital literacies "quite simply, involve the use of digital technologies for encoding and accessing texts by which we generate, communicate and negotiate meanings in socially recognizable ways" (p. 258). From a similar perspective, Martin (2005) proposes an expansive view of digital literacies as

> the awareness, attitude and ability of individuals to appropriately use digital tools and facilities to identify, access, manage, integrate, evaluate, analyze and synthesize digital resources, construct new knowledge, create media expressions, and communicate with others, in the context of specific life situations, in order to enable constructive social action; and to reflect on this process. (p. 135)

Much more than a set of skills or competencies, digital literacies are conceptualized as a way of being an engaged, responsible, reflective citizen in a 21st-century global community permeated by multimodal technologies. It is therefore critical that digital literacies be integrated throughout foreign/second language education, where multiple communities, identities, languages, and cultures converge.

2. Overview of the Volume

This volume opens with a chapter that reexamines the construct of communicative competence and proposes an upgrade for language learning and teaching in an era of new literacy practices that have emerged due to the continuously increasing access to and use of new digital tools and new types of communication spaces. Lotherington and Ronda reconceptualize the analogue framework originally proposed by Canale and Swain (1980) as a digital framework that includes four dimensions of competence: multimedia, collaborative communication, agentive participation, and multitasking. The authors of this chapter challenge teachers to ask themselves if curriculum, instruction, and assessment are truly preparing students for the realities of today's communicative landscape.

The next three chapters and the chapter by Peters and Frankoff offer survey-driven studies that use a variety of response formats and methods of analysis to investigate issues related to digital literacies. Williams, Abraham, and Bostelmann, for example, report the results of two large-scale surveys of undergraduate foreign language learners at a U.S. university. Their surveys include items that elicit information about ownership of digital devices, access to networked technologies, and new literacy practices. Although the students reported having relatively easy access to and high levels of familiarity with a wide range of digital tools and communication spaces, their opinions were almost evenly divided on whether or not social media should be integrated into the foreign language curriculum.

In this volume's second survey-driven study, Benini and Murray explore the terms *digital native* and *digital immigrant*, which were first proposed by Prensky (2001) as one way to categorize people according to whether or not—based solely on age—they possessed a presumed innate ability to understand all things digital. Through semistructured interviews, surveys, and classroom observations over a period of 18 weeks at two schools in Ireland, the authors were able to examine perceptions and uses of different types of technologies for academic and nonacademic purposes. At the end of their chapter, the authors recommend, among other things, avoiding the use of terms or labels (e.g., *digital native*, *digital immigrant*) that have no empirical basis. This is a sound recommendation applicable to many contexts, but this seems especially important in educational settings.

The chapter by Jiménez-Caicedo, Lozano, and Gómez uses Q methodology to measure students' subjectivities (i.e., shared values) related to the use of blogs as part of the Spanish curriculum. Three distinct perspectives emerged from the analysis of students' responses: one group perceived the use of blogs as a tool that could help them focus on language forms in order to improve their academic written Spanish (Perspective 1); another group viewed the blog project as a tool for communication, with attention to form being much less important (Perspective 2); and a third group considered blogs to be primarily a space for informal interaction (Perspective 3). This study reinforces the importance of determining clear expected learning outcomes, yet at the same time teachers must recognize that any tool can be used in a variety of ways by learners given the range of their attitudes, beliefs, and motivations.

The other survey-driven study in this volume explores the issue of plagiarism in the context of digital scrapbooking. Peters and Frankoff identify digital scrapbooking as a pedagogical task that may lend itself to plagiarism—perhaps unintentionally on the part of students—since it typically involves the use of small or large bits and pieces of other people's work that has been (re)arranged according to the goals of the project. The results of this study reveal, among other things, that teachers—in this case, at least—explain what plagiarism is to their students more frequently than they show students how to cite source material, a practice that can be somewhat difficult to understand if done infrequently.

In the next chapter, Kurek and Hauck propose a model for designing instruction grounded in multiliteracy, and they use task-based telecollaboration as the context for demonstrating this model, which has as its central components informed reception, thoughtful participation, and creative multimodal contribution. In this model, each of these components should be addressed at several levels: cognitive, social, discursive, and operational. The authors have proposed this model to close a digital divide related to the quality of online interactions rather than one that is related to the issue of access to networked technologies.

In a study of three international students at a U.S. university, Mina explores issues related to the ways in which multilingual learners represent themselves and enact their identities through multimodal narratives. The analysis of each case focuses on the structure (organization of shots and images) and content (visual, textual, and aural elements) of the students' multimedia projects. The author emphasizes the fact that there are different ways in which digital tools and literacy practices can be exploited to express cultural identities, and this process can be especially important for international students who might feel caught between or attempting to maintain some type of balance between different cultures and languages.

Reinhardt, Warner, and Lange report the results of a project designed to examine game-enhanced, literacies-informed instruction in a fifth-semester L2 German classroom at a U.S. university. Since one goal of this project was to promote the development of L2-mediated game literacies, participants engaged in gameplay activity as well as reading and writing about gaming discourses. Evaluations of the gaming unit by students produced mixed reactions, ranging from absolute enjoyment to dissatisfaction and frustration. Even though the results of this study reveal a substantial amount of tension between games, which are supposed to be fun, and learning, which is supposed to be serious, the authors point out that exploring the familiar and the contested can be a valid objective of L2 learning.

The next two chapters explore writing in the form of digital stories and digital social reading, respectively. Oskoz and Elola report the results of a case study in which learners of Spanish in an advanced writing course prepared a digital story over the course of a 16-week semester. Most of the students indicated that this project helped them to become better (L2) writers. Moreover, the authors note that creating digital stories allowed learners to move beyond traditional presentational modes of communication since they could also include images and sound, and they also had time for reflection since producing digital stories was a semester-long project. The chapter by Blyth offers four case studies of new literacy practices associated with e-reading devices, specifically eComma. Evaluations of digital social reading in two undergraduate courses and two graduate courses revealed several perceived affordances: coconstructed scaffolding, distribution of the cognitive load, synthesis of several activities into a single activity, group-level analysis, and blending various types of digital reading. Both of these chapters emphasize the interactive, social nature of digital literacies while exploring writing and reading—which were once considered individual activities—as collective endeavors.

The study by Mutta, Pelttari, Salmi, Chevalier, and Johansson investigates a range of information-seeking strategies of students in their L1 and L2 in order to better understand digital literacy practices and awareness in an academic setting. The findings from their study highlight the internet search engine as the dominant information-seeking strategy or tool, and the results demonstrate that instructional intervention can have an effect on information seeking, especially if awareness-raising tasks focus on the evaluation of the trustworthiness of resources (e.g., Wikipedia). The authors also note that learners can find the same information through many different paths and by using various tools in different ways.

After examining digital literacies in foreign and second language education from a variety of perspectives, the volume concludes with a chapter by Guikema and Menke that explores the role of digital literacies in teacher preparation. Their study analyzes teacher candidates' views of digital literacies in their current and future instructional practice and highlights the importance of explicitly addressing digital literacies in a cohesive way throughout teacher training.

3. Acknowledgments

Our sincere thanks go to all the reviewers who provided detailed and timely feedback: Kathleen Anderson de Miranda, Hélène Andrawiss-Dlamini, Isabel Alonso Belmonte, Melinda Dooly, Kate Douglass, María Fidalgo-Eick, Regine Hampel, Yuri Kumagai, Lina Lee, François Mangenot, Youngmin Park, Dolores Ramírez Verdugo, Jim Ranalli, Anita Saalfeld, Baburhan Uzum, and Rémi A. van Compernolle. We are also extremely grateful to the CALICO Monograph Series Editor, Robert Fischer, for the considerable time he devoted to this project and especially for his guidance and patience in its final stages.

References

Belz, J. A. (2002). Social dimensions of telecollaborative foreign language study. *Language Learning & Technology, 6,* 60–81.

Buckingham, D. (2008). Defining digital literacy: What do young people need to know about digital media? In C. Lankshear & M. Knobel (Eds.), *Digital literacies: Concepts, policies and practices* (pp. 73–89). New York, NY: Lang.

Byrnes, H., & Kord, S. (2001). Developing literacy and literary competence: Challenges for foreign language departments. In V. Scott & H. Tucker (Eds.), *SLA and the literature classroom: Fostering dialogues* (pp. 31–69). Boston, MA: Heinle & Heinle.

Canale, M., & Swain, M. (1980). Theoretical bases of communicative approaches to second language teaching and testing. *Applied Linguistics, 1,* 1–47.

Gee, J. P. (1999). Reading and the New Literacy Studies: Framing the National Academy of Sciences report on reading. *Journal of Literacy Research, 31,* 355-374

Kern, R. (2000). *Literacy and language teaching.* Oxford, England: Oxford University Press.

Kern, R., & Schultz, J. M. (2005). Beyond orality: Investigating literacy and the literary in second and foreign language instruction. *Modern Language Journal, 89,* 381–392.

Knobel, M., & Lankshear, C. (2008). Digital literacy and participation online social networking spaces. In C. Lankshear & M. Knobel (Eds.), *Digital literacies: Concepts, policies and practices* (pp. 249–279). New York, NY: Lang.

Kramsch, C., & Nolden, T. (1994). Redefining literacy in a foreign language. *Die Unterrichtspraxis/Teaching German, 27,* 28–35.

Lankshear, C., & Knobel, M. (2011). *New literacies: Everyday practices and classroom learning* (3rd ed.). New York, NY: Open University Press.

Martin, A. (2005). DigEuLit—a European framework for digital literacy: A progress report. *Journal of eLiteracy, 2,* 130–136.

Mercer, S. (2011). Understanding learner agency as a complex dynamic system. *System, 39,* 427–436.

New London Group. (1996). A pedagogy of multiliteracies: Designing social futures. *Harvard Educational Review, 66,* 60–93.

Pennycook, A. (2001). *Critical applied linguistics: A critical introduction.* Mahwah, NJ: Erlbaum.

Pool, C. (1997). A conversation with Paul Gilster. *Educational Leadership, 55,* 6–11.

Prensky, M. (2001). Digital natives, digital immigrants. Part 1. *On the Horizon, 9*(5), 1–6. Retrieved from http://www.marcprensky.com/writing/Prensky%20-%20Digital%20Natives,%20Digital%20Immigrants%20-%20Part1.pdf

Swaffar, J., & Arens, K. (2005). *Remapping the foreign language curriculum: An approach through multiple literacies.* New York, NY: Modern Language Association of America.

Chapter 1

2B or Not 2B? From Pencil to Multimodal Programming: New Frontiers in Communicative Competencies[1]

HEATHER LOTHERINGTON
York University (Canada)

NATALIA RONDA
Council of Ministers of Education (Canada)

Abstract

> Social communication has changed dramatically since Canale and Swain's (1980) statement of communicative competencies for French as a second language (FSL) testing, which outlined language/grammatical, sociolinguistic, discourse, and strategic competencies (see also Canale, 1983).[2] Communicative input/output was described in terms of four skills—reading writing, speaking, listening—that captured the media of communication of the times. However, times have changed, and communicative competence needs an upgrade to meet the needs of web 2.0 environments. This chapter reviews how communication practices have evolved over the past three decades in concert with developments in technical communications media and reconceptualizes the analogue framework of language/grammatical, sociolinguistic, discourse, and strategic competencies as a digital framework encapsulating multimedia, collaborative communication, agentive participation, and multitasking competencies.

1. Introduction

Communicative practices have gone through revolutionary changes since Canale and Swain's (1980) research statement of competencies to be tested in communicative approaches to French as a second language (FSL) instruction (see also Canale, 1983). Communicative competence comprised four components: language/grammatical, sociolinguistic, discourse, and strategic competencies de-

---

[1] This witty texting pun "2b or not 2b?" has been used by David Crystal (2008); see also Marshall and Moore (2013).
[2] See also Celce-Murcia, Dörnyei, and Thurrell (1995); Celce-Murcia (2007).

scribing what could be pedagogically expected of a learner of FSL in the context of Ontario, Canada. Communicative input/output was encapsulated as four skills: reading, writing, speaking, and listening, skills which captured the media of communication of the times. Tests were unquestionably written on paper with a pencil of specified graphite density (e.g., HB or 2B). Handwriting and typing on a typewriter were the processing mechanisms.

Fast forward three decades. ... Communicative competence is alive and well, the framework having been uncritically appropriated for communicative language teaching and absorbed as doctrine in communicative approaches to English as a second language (ESL) teaching (Leung, 2005; see also Savignon, 2007). The four-skills framework continues to form the backbone of gate-keeping English language tests, such as the Test of English as a Foreign Language (TOEFL), and the International English Language Testing System (IELTS), as exemplified in this quote from Educational Testing Services (ETS): "the TOEFL iBT® test is given in English and administered via the Internet. There are four sections (listening, reading, speaking and writing) which take a total of about four and a half hours to complete." (ETS, 2013). The Internet, in this quotation, is a vehicle for transmission of print text. However, the technical media of communication have changed dramatically, particularly over the past decade, deeply influencing how, where, when, why, and with whom we communicate, and the resultant communicative forms extend well beyond four identifiable skills.

Can a theory outlined for the communications media of the 1980s, then, still dependably encapsulate the communicative needs of today? This chapter focuses on how communication has changed over the past three decades; it questions the growing gap between social communication practices and formal language and literacy education, asking "what does communicative competence look like now?" Can we still reliably utilize a framework of language/grammatical, sociolinguistic, discourse, and strategic competencies?

2. Arriving at the Notion of Communicative Competence

> Languages are learned, not as forming in themselves a part of erudition or wisdom, but as being the means by which we may acquire knowledge and may impart it to others. (Comenius, 1657/1967, p. 203)

That languages are learned for the purpose of communication seems self-evident. Comenius clearly had a functional view of language teaching and learning in the 17th century. The formal learning of texts (in what were determined to be foreign languages) as they became publicly available following social uptake of the printing press, though, relied largely on grammar translation. Even Comenius' evidently multimodal approach to language learning (1659/1728), emphasizing practical sensory experience, promoted memorization of comparative texts (see Figure 1).

Figure 1
Picture Associations in Comenius, *Orbis Sensualium Pictus* (1659/1728)

Orbis Sensualium Pictus,

A World of Things Obvious to the Senses drawn in Pictures.

Invitation.	I.	Invitatio.
The Master and the Boy.		*Magister & Puer.*
M. Come, Boy, learn to be wise.		M. Veni, Puer, disce sapere.
P. What doth this mean, *to be wise?*		P. Quid hoc est, *Sapere?*
M. To understand rightly.		M. Intelligere recte,

Formal language instruction has historically focused on the grammatical structure of a language. Grammar translation, which comparatively attended to grammatical structure, as in Figure 1, was the principal approach to modern language learning in the 19th century (Stern, 1983) when the industrial revolution was paving the way for mass literacy and mass education. In fact, as Stern notes, in the 19th century, "grammar-translation was regarded an educationally valid mental discipline in its own right" (p. 454). Grammar translation, though, focused on literate language and generally ignored oral communication.

This lacuna was addressed in a reactionary manner by the direct method, which emerged in the second half of the 19th century. The direct method de-emphasized reading and writing, attending primarily to listening and speaking (Stern, 1983). By the mid 20th century, the audiolingual method, with roots in behaviorism and US military training, had gained prominence. Audiolingualism similarly parsed communication into oral and literate skills, which were to be learned in linear sequence, and focused on oral production (Stern).

The predominant approaches to formal (second and foreign) language teaching in the 19th and 20th centuries fit into Bosco and Di Pietro's (1970) typology of language instruction as *divergent* approaches: those separating language into skill areas, typically prioritized and sequenced. However grammar is approached in these methods—whether instrumental to language learning as in grammar translation or incidental as in audiolingualism, it is linguistically identified and linked to skills areas identified as *productive* (i.e., writing and speaking) and *receptive* (i.e., reading and listening).

Noam Chomsky's indelible footprint on the study of language acquisition moved linguistic structuralism to a different perspective wherein the speaker's competence was the basis for linguistic study. Language learning was not simply repetition and habit formation, a tenet of audiolingualism, but a hard-wired brain function capable of handling the elegance and complexity of grammatical acquisition. In Chomsky's (1965) abstract theorizing of language competence—identified as separate from performance in communication—grammatical structure formed the basis of linguistic competence.

Campbell and Wales (1970) and Hymes (1972) positioned social context as a governing force in human communication in direct, critical response to Chomsky's delimitation of linguistic competence as purely grammatical, promoting the notion of *communicative competence*, extending language competence from the human innateness ascribed by Chomsky into a social sphere. Canale and Swain (1980) applied the theoretical notion of communicative competence to the teaching of French as a second language in Canada, initially grappling with Chomsky's binary of *competence* and *performance*. By 1983, Canale had positioned communicative competence as "an essential part of actual communication but ... reflected only indirectly" (p. 5). Fifteen years later, Cazden (1997) argued that communicative competence was not reflected in performance, but learned from it.

3. Communication, Society, and Media Post-1983

Two major shifts have had a substantial effect on social, cultural, political, and economic life since the early 1980s: globalization; and digitization (see Figure 2).

Figure 2
Major Social Shifts: 1983-2013

Described as world shrinkage (Larsson, 2001) or flattening (Friedman, 2005), *globalization* has brought people and contexts together to form new paradigms of socializing, making connections, creating, and doing business. Parallel to globalizing trends is the increased focus on the local, reflected in the term *glocalization* (Robertson, 1995) which serves to remind that in the move towards the global, the local finds itself not only shaped by, but also shaping the global, emphasizing heterogeneity and diversity.

Closely aligned to glocalization is *superdiversity*. Vertovec (2007), who coined the term, described the superdiverse society in terms of "a dynamic interplay of variables among an increased number of new, small and scattered, multiple-origin, transnationally connected, socio-economically differentiated and legally stratified immigrants who have arrived over the last decade" (p. 1024). In 1980s Canada, the context in which Canale and Swain were working, cultural pluralism was framed as *multiculturalism* (Government of Canada, 2012). In the 21st century, Toronto is a superdiverse urban center in which students in local public schools speak over 75 languages (Toronto District School Board, 2014). In this dynamic cultural remix, individual repertoires and social practices alike are characterized by a mix of languages (see Figure 3).

Figure 3
Sign Board Outside Café in Toronto

Coste, Moore, and Zarate (2009) have reframed communicative competence in superdiverse contexts as *plurilingual* and *pluricultural competence* turning from language-based conceptualizations of cultural pluralism to a "focus on the individual as the locus and actor of contact" (p. v). In pilot projects across the greater

Toronto area, plurilingual practices are being edged into classroom conversations and textual products (see Cummins & Early, 2011; Lotherington, 2011).

Further to the conceptualization of plurilingual competencies is the place of machine-mediated collective intelligence whereby web-based programs and apps provide free translation assistance. Figure 4 depicts a Facebook exchange that utilized help from machine translation to contribute to a discussion about a song sung in Spanish with Portuguese subtitles generated by a German speaker.

Figure 4
Plurilingual Communication in Facebook Exchange

Heather Lotherington Muchas gracis, Susanne. Que linda cancion! Happy Thanksgiving, even though we Canadians celebrate this annual event in October. Mais c'et aujourd'hui pour les Americains. Vielen dank for the Spanish song with commentary translated into Portuguese. Wunderbar!
17 hours agp - like

Susanne   Hi Heather, ich bin sehr beeindruckt ☺!
See translation
16 hours ago - like

Arguably the most salient social shift of the past three decades has been *digitization*. The rapid expansion of digital technologies into social, economic, and cultural life has indelibly transformed communication practices. Revolutionary changes in communications media have wrought new tools and environments for communicating, which have, in turn, spawned new discourses, textual forms, and communities. The expansion of global digital networks together with the emergence of pocket-sized computer devices has enabled the individual to communicate across a global playing field of media access, production, and dissemination. Virtual ontologies have developed in cyberspace where people are involved in massively multiplayer worlds, such as Second Life (http://secondlife.com/), and massively multiplayer online role-playing games (MMORPG), such as World of Warcraft (http://us.blizzard.com/en-us/games/wow/), providing new identities and cultures of digital activity. There is even a virtual currency traded in real-time markets: bitcoin (http://bitcoin.org/).

The internet has spawned an era of fast capitalism (Agger, 1989, 2004) in which the boundaries between home and workplace have eroded. The evolution of the semantic web, or Web 2.0, has enabled greater social interactivity, inviting social media forums, such as Twitter (microblogging), Facebook (multimedia social profiling), and YouTube (video sharing)—sites and social networks involving digital technologies, media platforms, and communication skills that did not exist only a decade ago. Engaging with each other on line and off, we now constitute life in intersubjective realities (Brey, 2003a, 2003b).

From new neighbors to new social media, we have been swept into *convergence culture*. Web 2.0 has provided a petri dish for new ways of communicating, enabling collaborative authorship structures (e.g., wikis), multimodal communication environments (e.g., YouTube), and chatting in unicode (e.g., texting). It has destabilized binary distinctions grounding 1980s language education, such as first and second (or foreign) language and reader/writer (Kramsch, 2009). Competent communication cannot with any validity be separated into isolated skills, relegated to unmediated single authorship or limited to predigital domains.

4. New Frontiers in Communication

> It is not always clear just what skills are included in theories of basic communication skills. (Canale & Swain, 1980, p. 9)

The dyadic split between communicative competence and performance has faded in late modernity where the rapidly spinning engineering cycle of technical media, profusion and choice in social media platforms, and access to near-ubiquitous connection encourage jumping into the middle of a communicative activity—be it gaming, blogging, texting, Facebook posting—and figuring it out during rather than before participation: a postmodern performance-before-competence approach wherein competence is illusory, given the complexity and lack of fixity of technical media. The blinding profusion of technical media in perpetual update, discovered in social practice, and purchased on online stores offering products, upgrades, technical help, and interactive forums has changed who uses what, where, when, how, and why to say or do or be what. So much for pencil grip, handwriting, remembering individually how to spell a word, and forming thoughts logically and grammatically into static textual forms. Chatting in digital forums does not need oral speech; reading and writing in social media platforms require read/write programming; a text may be dynamic, multimodal, remixed, and collaboratively created; crowd-sourced digitally available linguistic, grammatical, and spelling help is available online, and even coded into sites and software.

How does one begin to teach language in this new digital media environment?

4.1 Media and Mediation

> In a culture like ours, long accustomed to splitting and dividing all things as a means of control, it is sometimes a bit of a shock to be reminded that, in operational and practical fact, the medium is the message. This is merely to say that the personal and social consequences of any medium—that is, of any extension of ourselves—result from the new scale that is introduced into our affairs by each extension of ourselves, or by any new technology. (McLuhan, 1965, p. 7)

Mediation is fundamental to communication. This puts into question whether communicative competence is truly measurable by psychometric means, such as standardized language testing. Certainly in the digital communication environments of today, communicative competence is distributed—not only among interlocutors, but also with machine mediation, a point Rushkoff (2011) compellingly brings home:

> Our screens are the windows through which we experience, organize and interpret the world in which we live. We are doing more than extending human

agency through a new linguistic or communications system. We are replicating the very function of cognition with external, extra-human mechanisms. These tools are not mere extensions of the will of some individual or group, but entities that have the ability to think and operate other components in the neural network—namely, us." (pp. 31–32)

The term *medium* has a very broad reach, describing different aspects of communication: languages (e.g., French as the medium of instruction), textual surfaces (e.g., paper), tools of textual production (e.g., ballpoint pen), and means of transmission (e.g., public broadcasting). *New media* takes in a very large swath, from technological interface to media genre.

In Elleström's (2010) *intermediality* paradigm, media are categorized into three types:

- *basic* (defined by four modal properties: material, sensorial, spatio-temporal, semiotic);
- *qualified* (aesthetic and communicative properties); and
- *technical* (the display or realization mechanism).

Elleström (2010) explains:

Basic media are simply defined by their modal properties whereas qualified media are also characterized by historical, cultural, social, aesthetic and communicative facets. Technical media are any objects, or bodies, that 'realize', 'mediate' or 'display' basic and qualified media. (p. 5)

Over the past three decades there have been massive changes in technical media, which deeply affect basic and qualified media. The predominant technical media of 20th-century learning, namely pencils, ballpoint pens, paper, and books, relied on orthographic encoding in the (Roman) alphabet, a nonmaterial technology (Faulkner & Runde, 2011). Such fundamental 20th-century literacy tools have been augmented and superseded by pocket-sized, networked digital technologies that utilize a complex digital toolkit claiming the pixel as base unit (Cope, Kalantzis, & Lankshear, 2005). This profound technical media shift has changed the very essence of literate communication as literacy has shifted from page to screen (Kress, 2003, 2010; Snyder, 1997, 2002) and from reading and writing alphabetically to multimodal programming.

Internet connection is now near-ubiquitous in many parts of the world, and disruptions to online connections are considered annoying aberrations in day-to-day functioning. Moreover, technical communications media have become miniaturized to the point where they are worn as apparel, creating a cyborg interface in communication. Bluetooth attachments for smart phones fit inside or over ears or clip onto lapels, MP3 players slot into armbands and belt loops, and gadgets from lifelogging cameras to fitness monitors clip onto pockets or slide over wrists. These devices facilitate a hybridized online-offline existence and enable learners to capture the very scenarios they wish to understand and interact within.

4.2 Technology

Conceptually, technology is a Trojan horse, encapsulating knowledge and systems, social and economic activity, and physical artifacts (Bijker, 2010; Lawson, 2010). A ballpoint pen, for instance, is a communications technology. With this in mind, we see that the four skills—reading, writing, listening, speaking—that characterized communication with 1980s technology are inadequate to describe contemporary practices that capitalize on new media affordances, creating new conventions, discourses, and communities, and requiring new practices that are being played out in day-to-day experimental usage. Theoretical revision of what it means to be communicatively competent in dynamic, web-based communication environments is, in digital parlance, overdue for a critical update.

4.3 Technological Extension: Multimedia Communication

As McLuhan (1965) famously foretold, technologies evolve from extending our abilities to become extensions of our consciousness:

> After more than a century of electric technology, we have extended our central nervous system itself in a global embrace, abolishing both space and time as far as our planet is concerned. Rapidly, we approach the final phase of the extensions of man—the technological simulation of consciousness, when the creative process of knowing will be collectively and corporately extended to the whole of human society … . (pp. 3–4)

Digital technologies have radicalized our thinking about what is possible: our senses are augmented, our consciousness is extended, and our very social fabric is constantly reimagined. It is a radically new way of thinking about the world that digital technologies introduce: a world where information can be broken down, manipulated, and stored virtually; where we transcend physical barriers to play, work, and create in a digital environment. Digital media enable interactive multimediality: multimodality and multidimensionality cut across physical and geographical borders to bring rich content and connect people in different parts of the globe. Digital multimediality captures and augments all senses: through moving and still images, sound, and movement, augmented and responsive environments are created, where human and machine become one. Beyond the extended capabilities to manipulate information and objects digitally, the substance of what we can do has changed. Multimediality has changed the patterns of human interaction to an unprecedented extent: collaborative possibilities through and with digital media are virtually endless.

Digital multimediality presents the most significant challenge to the understanding of communication. The boundaries of what can be understood as speech are blurred: speech can be immediate or recorded, typed, texted or spoken, human or machine generated. The speech event is reframed across multimedia as a 140-character tweet, a YouTube video, a comment on a message board, a MMPORPG play, a snippet of computer code, or a multimedia lecture. The minimal unit of communication has shrunk to the bit (Negroponte, 1995), yet the variety of communication spawned by the bit is seemingly limitless.

Digital communication certainly extends semiotic possibilities beyond the traditional walls of language: different media and modalities are spliced to create new meanings. This is not optional extension: the digital citizen is versed in digital genres that wrap language into multimedia expression. In order to be an effective communicator in digital environments, one must be able to navigate a complex web of shifting meanings created through multiple media, as well as become adept at splicing these meanings in personal production. The simple activity of maintaining a social networking account, for instance, involves such novel and complex skills as manipulating digital audio and visual materials, typed language, and novel programming behaviors, such as friending, liking, and sharing.

4.4 Speakers as Agentive Users

> We've made huge changes in the way we communicate and engage with each other, thanks to the Internet. We're back in touch with high school buddies and long-lost relatives, and we can get information on just about anything and say just about anything we want online. (Cross, 2011, pp. 135–136)

What sets apart the digital speaker? Crucially, digital media have given users unprecedented opportunities for global collaboration. Though digital participation is still predicated on access to digital technologies and facilitated by a knowledge of English, the importance of English as the digital lingua franca has diminished with recent waves of cyberglocalization (Morbey, 2006) engaging localized digital communities.

New digital competencies essentially include collaboration, with social, gaming, and professional platforms offering different patterns of working together. The concept of expert knowledge is no longer equated with individual knowledge. As more projects solicit and become reliant on input from multiple users across the globe, the notion of collective digital problem solving becomes more real. Crowdsourcing, or putting minds together to create a massive problem-solving collective, has been gaining momentum in business and science. As Davidson (2012) suggests, digital models of collaboration like crowdsourcing suggest a drastically new model of problem solving: incrementally tackling challenges, and being agile in approaching the problem at hand.

Digital media offer a new model of approaching content altogether, one that presents a great challenge for the classroom. Skill-based roles of speaker-listener-reader-writer do not make sense in digital domains where users are agents: creators as well as consumers of content with knowledge of pragmatic media access of more utility than ability to manipulate user-generated language. From Facebook and YouTube to computer gaming, users interact with digital platforms and with each other in the production of content that ranges from simple text messages to elaborate multimedia productions. This content is not being marked for correctness, but rather taken on its merit and contribution, potentially by a large and diverse audience.

5. Communicative Competencies Reimagined

> The struggle now is to find new metaphors and constructs that would capture multilingual communication. How do we practice a linguistics that treats human agency, diversity, indeterminacy, and multimodality as the norm? (Canagarajah, 2007, p. 935)

What counts as communicative competence must be rethought to match the shifting landscape of what it means to be an effective communicator in the digitally connected world. We need to reconceptualize the analog framework of language/grammatical, sociolinguistic, discourse, and strategic competencies as a digital framework; we propose as a starting gambit: multimedia, collaborative communication, agentive participation, and multitasking competencies.

Table 1 sketches how shifting technological and social trends might play out in reimagining communicative competence or, to borrow from technological parlance, communicative competence 2.0.

Table 1
Communicative Competence 2.0

Trend	Communicative competence 2.0
Multimedia competency	• Navigating multiple digital platforms • Utilizing multiple modalities for meaning making • Coding competence: human and machine languages • Remediating old media to produce new meanings
Collaborative communication	• Collaborative meaning production, including multiple authorship • Ongoing learning through critical dialogue and engagement • Recognizing and harnessing machine mediation • Social participation in global forums
Agentive participation	• Learning by doing: performance towards competence • Accessing, creating, sharing, remixing purposeful content • Authority reimagined • Accessing and joining sites of interest and involvement
Multitasking	• Managing multiple sites and activities • Navigating breadth and depth of engagement

If language teaching is to borrow a page from the digital revolution, it is this: digital users are recast from novice to expert in the making. Learners are agentive in their own learning, not bereft of authority as in the analog classroom. Linguistic form is supplemented in agentive ways by employing complementary meaning-making channels in textual production attentively participating in social communication, and consulting collective machine-based intelligence, such as help desks.

In agentive participation, performance leads to competence, as Cazden hypothesized in 1997. Learning happens in situ and users learn as they go along, being thrown into the depth of new media and allowed to experiment, explore, fail, and reach out to their digital community of support to provide guidance. Users are not marked on progress in separate skill areas; rather, they are judged on the success of their participation and production, for example, a poignant social commentary in a Facebook post, a masterfully remixed video posted on YouTube, or a digitally retouched photo on Instagram. Each user has a chance to be a part of an informal learning community in which expert and nonexpert roles are blurred and success is judged on merit, not a decontextualized metric.

Communicative competence reconceptualized for the language learner must take account of this changing paradigm of participation, and reflect the idiosyncratic and varied patterns with which digital media allow us to communicate and become experts at using them. Communicative competence 2.0 captures this idiosyncrasy by taking account of the diversity of digital competencies required for successful communication, and by acknowledging that learners enter into digital communicative domains with various levels of comfort, taking varied and self-guided paths towards expertise.

Communicative competence digitally rethought also stresses the role of peers in supporting the development of expertise. Digital learning is less hierarchical, more flat and dispersed among a broad community of users with varied levels of expertise. In a sense, Illich's (1971) vision for webs of learning is realized online: novices rely on the experience of veterans to engage in self-guided informal learning. Competence is formed in critical dialogue with peers, which includes feedback, advice, banter, and encouragement.

The incredible diversity and abundance of digital building blocks complicate digital communication more than ever. The traditional media of communication have been augmented by digital media, offering users unprecedented opportunity to splice, remix, and produce meaning digitally. New conventions of communicative media offer an additional challenge: users need to find media-appropriate and rich ways to communicate, combining digital media and modalities in novel ways. This diversification of the communicative repertoire means that communicative competence 2.0 is adaptive and shifting, responsive to changes in the communicative landscape. Fundamental to managing this complexity is multitasking.

Globalization and glocalization pressures shape communicative competence by reducing the importance of a human language as a lingua franca. It could be argued that computer code, the lifeblood of digital communication, is the new lingua franca. Indeed, the communicative mediation inherent in computer coding offers users new and unprecedented opportunities not only to participate, but also to construct digital communicative platforms. Lack of basic understanding of the significant place of machine mediation is an impediment to using the potential of networks for learning. Through open online learning platforms, hundreds of thousands of people have the opportunity to learn programming. *Learn to program: The Fundamentals* is Coursera's third most popular course by number of total enrollments with close to 200,000 unique learners (Coursera, 2013). This trend is emblematic of a new communicative competence that is steadily emerging—an agility with the code that underscores all digital communication.

Digital media have opened unprecedented opportunities for an overhaul of what communicative competence means. Collaborative possibilities, the erosion of the expert, delisting of the native speaker and enlisting of the social participant, the growing importance of programming languages, and the need to produce meaningful and powerful media all contribute to a renewed understanding of what it means to be an effective communicator in the 21st century.

How can we assemble and develop all of this for classroom language teaching?

6. Communicative Competence 2.0 in the Language Classroom

Communicative competence 2.0 in the language classroom is reimagined through three lenses: collaborative communication and authorship; agentive participation and authority; and multimedia competency and authenticity. These critical lenses capture changes to the communicative landscape introduced by digital technology, and bring into focus the practical dimensions of communicative competence 2.0 reimagined for today's effective communicator.

6.1 Multimedia Competency: Authenticity is Learner Driven

Multimedia competency is at the crux of communicative competence reimagined, but what does it mean to have authentic multimedia-driven communication, and what might it look like in the classroom? Authentic communication in the digital world is made not born. There is no speaker native to the virtual world, and the concept of digital native (Prensky, 2001), which assigns innate expertise by birthright, is both dated and ageist. Digital competencies are socially learned, not innately developed. The user engages with audiences around the world in order to seek, create, entertain, and find solutions to pressing problems. The range of communication may vary from creating or following memes[3] (e.g., the Japanese YouTube sensation Maru the cat who became famous for his particular affinity for jumping into boxes; http://bit.ly/jd9CFJ) to participating in a crowd-sourced problem-solving team effort to find a cure for HIV/AIDS. But one thing is true: users engage in tasks and contexts that are purposeful; in this their engagement is unique, diverse, and authentic.

Language classrooms need to welcome multiple digital platforms to enable environments authentic to the participating learner: Twitter as a platform for keeping a microjournal, Facebook as a class social networking tool, and YouTube as a video logging opportunity — virtually any digital platform can lend itself to use in the language classroom. The multitude of media and modalities can be authentically tapped (e.g., text, image, sound), and the more complex modalities of video and audio editing, interactive media, and touch-based technologies can significantly augment language development.

Digital platforms are multiplying every day; some are global in scope, and others are glocal (i.e., targeted towards a specific linguistic or cultural group with worldwide membership). Research with a group of multilingual teenagers (Sinits-

---

[3] About.com defines 'meme' as "a virally-transmitted cultural symbol or social idea" (http://abt.cm/1iTmJcs).

kaya Ronda, 2011) indicated that they were discerning digital consumers engaged in digital sampling as communicative practice: trying out various digital tools and picking those found to be most useful to their purposes. In a similar fashion, a language classroom can engage with these sampling practices, tapping into the learners' digital repertoires and finding a fit between learners' digital practices and learning objectives.

6.2 Collaborative Communication: New Authorship Models

The shifting of authorial practices from single to multiple authors and the emergence of collaborative authorial practices have significant implications for the understanding of communicative practices in the classroom. Shared authorship has game-changing implications for teaching, learning, and, particularly, assessment, all of which are predicated on individual accomplishment in unnatural isolation from the learning community. The power of collaborative digital authorship has transformed the workplace: people routinely collaborate with colleagues across the country, and entire projects come together without collaborators ever meeting each other face to face. Socially, remotely connected people engage in communal research (e.g., on Wikipedia), writing (e.g., on fan fiction sites), gameplay (e.g., on MMPORPGs), and even virtual economic pursuits (e.g., in Second Life).

Parallel with this shift towards collective authorship is a growing space for the individual voice: never before in history has there been such an abundance of self-publishing platforms and opportunities. Anyone with a computer and an internet connection can make a personal opinion public. The rise of digital citizenship journalism has brought the power of the individual into the spotlight, as in the role of Twitter in the Arab Spring uprisings (Harlow & Johnson, 2011). In the classroom, individual digital production can be a powerful tool, and the digital content production that learners may already be engaging in on a daily basis can and should be incorporated into formal learning.

In the classroom, shifting authorial practices need to be reflected in collaborative peer work which can span from face-to-face to virtual contexts and extend from the classroom to the global community. Collaborative work can be facilitated through peer sharing and peer feedback, collaborative composition and media creation, and participation in virtual class discussions. More important, collaboration can happen outside the classroom with the wider community of users in the production of digital content, allowing learners to engage with cutting-edge digital media and make a contribution to a current event.

6.3 Agentive Participation: Sharing Authority

Who creates content in the classroom?

As novice users take charge of their learning with new media, they self-organize into informal communities of authors. With shared authorship comes collective authority, which calls into question the dimension of content production. This has heavy implications for the classroom given the traditional top-down role of the teacher. As we theorize in this chapter, digital media have allowed for the emergence of a grassroots expertise that is created out of interest, or necessity, and is acquired by learners in iterative cycles involving practice, learning, and discussion with a community of experts. The grassroots expertise that learners bring into the classroom can be harnessed to create a learning environment where learners create meaningful and immediate digital content. This content can then be shared with peers in the classroom and the global community.

Digital publications have come under critique from educators for their perceived lack of robustness and authority that is so valued in the formal classroom. However perhaps this quality, in fact, makes these media powerful learning tools. On a platform like Facebook, YouTube, or Wikipedia, where no user is a judgmental authority, true collaborative content production can happen. Learners can engage in nonhierarchical free play with new media and allow themselves to be creative, critical, and even audacious. On the internet nobody knows you are a learner: expertise is granted through participation and experimentation.

Another dimension of authority relates pointedly to assessment: How can communicative competence be assessed in a climate of dynamic, shared authorship? Digital revamping of communicative competence calls into question analog assessment mechanisms. In order to bring assessment up to date, we need to look to existing paradigms of how digital participation and production is judged. The grassroots nature of digital participation brings to the forefront the importance of peer assessment: digital content is liked, shared, retweeted, remixed, and distributed within a community of users. Peer feedback is crucial to the development of digital skills and honing of expertise, and it can be harnessed in the classroom. Developing metrics that take into account learners' individual and collaborative performance will be a challenge, but this is a necessary step toward updating a classroom practice that relies heavily on contrived assessment constructs. Collaborative assessment could include as part of the suite of assessment approaches a peer-assessment component that mirrors the feedback commonly built into social media platforms.

6.4 Multitasking: Navigating Complexity

It is important to recognize that communicative competence 2.0 is characterized by multitasking. With the unprecedented level of media and modalities, the success in navigating digital waters depends greatly on being able to keep multiple goals in sight and keeping track of multiple strands of digital conversations. It is common for digital users to engage in a host of digital activities simultaneously (see Figure 5), such as keeping a social profile on Facebook, including checking the news and commenting on the articles posted; tweeting and retweeting; and keeping a log of social bookmarks via Pinterest. Communicative competences work in concert rather than discretely: to be a proficient digital user moving with the pace of the digital traffic, one must acquire the strategic skills to competently navigate complex and interacting media, platforms, files, and communicative purposes.

Figure 5
Digital Multitasking

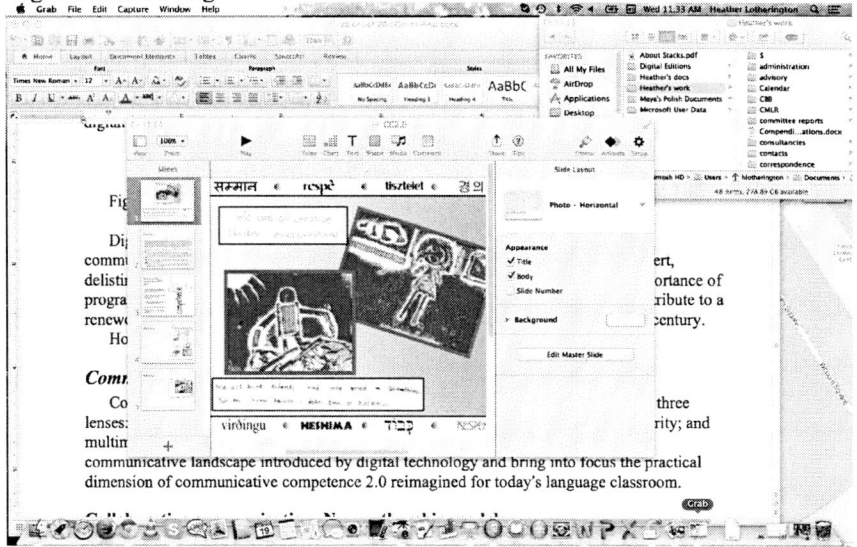

7. 2B or Not 2B: From the Pencil to the Program

> I am here to declare the enemy: the number 2 pencil. (Hayes Jacobs, 2011, 0.36–0.42)

Over the past few decades, communication and society have changed substantially. Trends towards globalization are evident in glocalized superdiverse urban populations, increasingly multilingual landscapes, and recognition of individuals' plurilingual repertoires. Into the global has come a glocal consciousness, taking the neighborhood to the world and bringing the world to the local community.

The digital revolution, so to speak, has ousted the industrial revolution as the driving force in social and economic life and, arguably, in cultural life as individuals merge their online-offline social lives. The media of communication—from technical media to basic and qualitative aspects of mediation—have wrought an immensely complicated multimedia landscape within which we now communicate on a daily basis. The technological devices we use to communicate have shrunk in size as they have increased in power. Where once management of a pencil was sufficient to encode daily communication, now multitasking across multimedia is a basic.

In this massively changing communicative landscape, we must question our roles as teachers of languages. Language is still the predominant semiotic resource available to us as social beings. But the technological media for encoding and decoding languages have changed how this is done. The elemental fours skills toolkit taught for centuries is too flat, incapable of handling merged, combined, augmented, linked, animated, orchestrated, remixed, and reimagined communication processes. The classroom, and particularly the testing culture that is steering far too much of classroom language teaching, still clings to four skills and notions of communicative competence fossilized in pencil and paper technologies. We need to question what social world we are teaching to.

Seventy-five years ago, Professor Harold Benjamin (Peddiwell, 1939/2004) pseudonymously published a parody of education's failure to keep abreast of environmental change in which the elders maintained an ice age curriculum that was tried and trusted despite its irrelevance to the life-and-death realities of the new environment. In language teaching and language assessment today, we are staring at a communicative ice age where the affordances of pencil and paper have been transposed uncritically to new media environments. Communicative competence has changed, and changed dramatically. We cannot merely jam multimedia into grammatical competence or collaborative communication into discourse competence. The social is a participatory landscape wherein the learner is highly agentive: a multitasker whose authentic needs can be captured for classroom sharing — just as they are in social media sites now.

Communicative competence needs an upgrade to 2.0 environments. The stakes are huge: we need to ask boldly, are we preparing students, young and old, for the world in which they need to communicate, indeed the world which they will access, and share in creating? This chapter has offered a start-up polylogue on how we are communicating and what we need to do to upgrade language teaching basics for new media environments.

References

Agger, B. (1989). *Fast capitalism: A critical theory of significance*. Urbana, IL: University of Illinois Press.

Agger, B. (2004). *Speeding up fast capitalism: Cultures, schools, families, jobs, bodies*. Boulder, CO: Paradigm.

Bijker, W. E. (2010). How is technology made?—That is the question! *Cambridge Journal of Economics, 34*, 63–76.

Bosco, F. J., & Di Pietro, R. J. (1970). Instructional strategies: Their psychological and linguistic bases. *International Review of Applied Linguistics, 8*, 1–19.

Brey, P. (2003a). Design and the social ontology of virtual worlds. In D. Koepsell & L. Moss (Eds.), *John Searle's ideas about social reality: Extensions, criticisms and reconstructions* (269–282). Malden, MA: Blackwell.

Brey, P. (2003b). Theorizing modernity and technology. In T. J. Misa, P. Brey, & A. Feenberg (Eds.), *Modernity and technology* (pp. 33–71). Cambridge, MA: MIT Press.

Campbell, R., & Wales, R. (1970). The study of language acquisition. In J. Lyons (Ed.), *New horizons in linguistics* (pp. 242–260). Harmondsworth, England: Penguin.

Canagarajah, S. (2007). Lingua franca English, multilingual communities, and language acquisition. *Modern Language Journal, 91*, 923–939.

Canale, M. (1983). From communicative competence to communicative language pedagogy. In J. C. Richards & R. W. Schmitt (Eds.), *Language and communication* (pp. 2–27). London, England: Longman.

Canale, M., & Swain, M. (1980). Theoretical bases of communicative approaches to second language teaching and testing. *Applied Linguistics, 1*, 1–47.

Cazden, C. B. (1997). Performance before competence: Assistance to child discourse in the zone of proximal development. In M. Cole, Y. Engestrom, & O. Vasquez (Eds.), *Mind, culture and activity: Seminal papers from the laboratory of comparative human cognition* (pp. 303–310). Cambridge, England: Cambridge University Press.

Celce-Murcia, M. (2007). Rethinking the role of communicative competence in language teaching. In E. Alcón Soler & M. P. Safont Jordà (Eds.), *Intercultural language use and language learning* (pp. 41–57). Dordrecht, The Netherlands: Springer.

Celce-Murcia, M., Dörnyei, Z., & Thurrell, S. (1995). Communicative competence: A pedagogically motivated model with content specifications. *Issues in Applied Linguistics, 6*(2), 5–35.

Comenius, J. A. (1967). *The great didactica* [Didactica Magna]. (M. W. Keatinge, Trans.). New York, NY: Russell & Russell.

Comenius, J. A. (1728). *Orbis Sensualium Pictis*. (C. Hoole, Trans.) London, England: Sprint.

Chomsky, N. (1965). *Aspects of the theory of syntax*. Cambridge, MA: MIT Press.

Cope, B., Kalantzis, M., & Lankshear, C. (2005). A contemporary project: An interview. *E-learning and Digital Media, 2*, 192–207.

Coste, D., Moore, D., & Zarate, G. (2009). *Plurilingual and pluricultural competence: Studies towards a Common European Framework of Reference for language learning and teaching*. Strasbourg, France: Council of Europe. Retrieved from http://www.coe.int/t/dg4/linguistic/Source/Source Publications/CompetencePlurilingue09web_en.pdf

Coursera, Inc. (2013). *Top 9 courses on Coursera by lifetime enrollment*. Retrieved from https://twiter.com/coursera/status/379650980154863617

Cross, M. (2011). *Bloggerati, Twitterati: How blogs and Twitter are transforming popular culture*. Santa Barbara, CA: Praeger.

Crystal, D. (2008, July 5). 2b or not 2b? *The Guardian*. Retrieved from http://www.theguardian.com/books/2008/jul/05/saturdayreviewsfeatres.guardianreview

Cummins, J., & Early, M. (Eds.). (2011). *Identity texts: The collaborative creation of power in multilingual schools*. Stoke-on-Trent, England: Trentham Books.

Davidson, K. (2011). *Now you see it: How technology and brain science will transform schools and business for the 21st century*. New York, NY: Penguin Books.

Educational Testing Services. (2013). *TOEFL iBTFL iBT content*. Retrieved from http://www.ets.org/toefl/ibt/about/content/

Elleström, L. (2010). The modalities of media: A model for understanding intermedial relations. In L. Elleström (Ed.), *Media borders, multimidality and intermediality* (pp. 11–48). Houndmills, England: Palgrave Macmillan.

Faulkner, P., & Runde, J. (2011). *The social, the material, and the ontology of non-material technological objects*. Paper presented at the 27th EGOS Colloquium, Gothenburg, Sweden.

Friedman, T. (2005). *The world is flat: A brief history of the twenty-first century*. New York, NY: Farrar, Straus & Giroux.

Government of Canada (2012). *Canadian multiculturalism: An inclusive citizenship*. Retrieved from http://www.cic.gc.ca/english/multiculturalism/citizenship.asp

Harlow, S., & Johnson, T. J. (2011). Overthrowing the protest paradigm? How the *New York Times*, *Global Voices* and Twitter covered the Egyptian revolution. *International Journal of Communication, 5*, 1359–1374.

Hayes Jacobs, H. (2011). *TEDxNYED*. Retrieved from http://bit.ly/O90CU

Hymes, D. (1972). On communicative competence. In J. B. Pride & J. Holmes (Eds.), *Sociolinguistics* (pp. 5373). Harmondsworth, England: Penguin Books.

Illich, I. (1971). *Deschooling society*. New York, NY: Harper & Row.

Kramsch, C. (2009). Third culture and language education. In L. Wei & V. Cook (Eds.), *Contemporary applied linguistics Vol. 1: Language teaching and learning* (pp. 233–254). London, England: Continuum.

Kress, G. (2003). *Literacy in the new media age*. London, England: Routledge.

Kress, G. (2010). *Multimodality: A social semiotic approach to contemporary communication*. Abingdon, England: Routledge.

Larsson, T. (2001). *The race to the top: The real story of globalization*. Washington, DC: CATO Institute.

Lawson, C. (2010). Technology and the extension of human capabilities. *Journal for the Theory of Social Behaviour, 40*, 207–223.

Leung, C. (2005). Convivial communication: Recontextualizing communicative competence. *International Journal of Applied Linguistics, 15*, 119–144.

Lotherington, H. (2011). *Pedagogy of multiliteracies: Rewriting Goldilocks*. New York, N.Y: Routledge.

Marshall, S., & Moore, D. (2013). 2b or not 2b plurilingual? Navigating languages, literacies, and plurilingual competence in postsecondary education in Canada. *TESOL Quarterly, 47*, 472–499.

McLuhan, M. (1965). *Understanding media: The extensions of man.* New York, NY: McGraw-Hill.

Morbey, M. L. (2006). Killing a culture softly: Corporate partnership with a Russian museum. *Museum Management and Curatorship, 21*, 267–282.

Negroponte, N. (1995). *Being digital.* New York, NY: Vintage Books.

Peddiwell, J. A. (2004). *The sabre-tooth curriculum.* New York, NY: McGraw-Hill.

Prensky, M. (2001). Digital natives, digital immigrants. Part 1. *On the Horizon, 9*(5), 1–6.

Robertson, R. (1995) Glocalization: Time-space and homogeneity-heterogeneity. In M. Featherstone, S. Lash, & R. Robertson (Eds.), *Global Modernities* (pp. 25–44). London, England: Sage.

Rushkoff, D. (2011). We interrupt this program. *School Library Journal, 57*, 30–32.

Savignon, S. (2007). Beyond communicative language teaching: What's ahead? *Journal of Pragmatics, 39*, 207–220.

Sinitskaya Ronda, N. (2011). *Facing the Facebook challenge: Designing online social networking environments for literacy development* (Unpublished doctoral dissertation). York University, Toronto, Ontario, Canada.

Snyder, I. (Ed.). (1997). *Page to screen: Taking literacy into the electronic era.* St. Leonards, New South Wales, Australia: Allen & Unwin.

Snyder, I. (Ed.). (2002). *Silicon literacies: Communication, innovation and education in the electronic age.* London, England: Routledge.

Stern, H. H. (1983). *Fundamental concepts of language teaching.* Oxford, England: Oxford University Press.

Toronto District School Board. (2014). *About us: Quick facts.* Retrieved from http://www.tdsb.on.ca/AboutUs/QuickFacts.aspx

Vertovec, S. (2007). Super-diversity and its implications. *Ethnic and racial studies, 30*, 1024–1054.

Chapter 2

A Survey-Driven Study of the Use of Digital Tools for Language Learning and Teaching

Lawrence Williams
University of North Texas (USA)

Lee B. Abraham
Columbia University (USA)

Evan D. Bostelmann
University of Wisconsin at Milwaukee (USA)

Abstract

The present study reports the results of two large-scale surveys of ways in which undergraduate foreign language learners perceive and use digital tools. These two surveys, which were both administered during 2013, represent the first part of a long-term project that aims to provide faculty and administrators with a more complete view of undergraduate students' preferences and practices related to new technologies and digital literacies.[1] The goal of the project is to gain a better understanding of how new digital tools are being used in order to make informed decisions about the future of the foreign language curriculum, faculty and student training, and other related issues. The results suggest that young adults do not constitute a monolithic group of people who can be easily categorized as so-called digital natives since a variety of factors can influence students' digital literacy practices as well as their access to and use of digital tools.

1. Introduction

The rapid creation and diffusion of technologies continually redefine what it means to be able to engage in communicative activity in our social and profes-

---

[1] During 2013, the first phase of the project focused on developing data collection instruments in the form of surveys. During 2014, the second phase of the project is using survey results as topics for semistructured interviews in the form of focus groups (i.e., more in-depth answers, but with fewer participants). During 2015, the final phase of the project, results from the previous two phasest will be used to develop a refined large-scale survey.

sional lives. As Leu, Kinzer, Coiro, Castek, and Henry (2013) observe, "to be literate tomorrow will be defined by even newer technologies that have yet to appear and even newer discourses and social practices that will be created to meet future needs" (p. 1150). Such circumstances pose unique opportunities for language educators and administrators to effectively integrate technologies in a world in which "communicative competence requires mastering new literacies: acquiring grammars of visual and multimodal design for reading and writing hybrid text types ... managing tools for consuming and producing multimodal texts, and mastering changing discourses within work environments and key social institutions" (Lankshear, Knobel, & Curran, 2013, p. 864).

Several scholars have attempted to define the characteristics of adolescents and young adults who increasingly use information and communication technologies (ICTs). Prensky (2001) was one of the first to propose a definition of the generation born and raised with ICTs who "are all 'native speakers' of the digital language of computers, video games and the Internet" (p. 3).[2] According to Prensky:

> Digital Natives are used to receiving information really fast. They like to parallel process and multi-task. They prefer their graphics before their text rather than the opposite. They prefer random access (like hypertext). They function best when networked. They thrive on instant gratification and frequent rewards. They prefer games to "serious" work. (2001, p. 2)[3]

Presumably, they are all technologically savvy, and they all rely heavily on emerging technologies for everyday communications and tasks. Conversely, their parents and teachers are viewed as digital immigrants who were born and raised in a predigital age during which many of these new technologies were not available. Some observers consider digital natives to be members of a "wired generation" with a seemingly innate knowledge base that allows them to use a rather sophisticated skill set to use digital tools with ease (see Palfrey & Gasser, 2008; Rosen, Carrier, & Cheever 2010; Tapscott, 2009). Although members of the Net Generation may indeed use digital media for a considerable portion of their daily activities, recent studies suggest that such use does not necessarily indicate that these so-called digital natives are always able to complete academic assignments or real-world tasks effectively with ICTs (e.g., Hubbard, 2013; Junco, 2012; Rosen, Carrier, & Cheever, 2013).

This chapter addresses the need for research on language learners' adoption and use of new technologies primarily for academic purposes. One catalyst for the present study, which is part of a larger computer-assisted language learning (CALL) project examining the use of new digital tools, is a recent article (Smith,

---

[2] For a discussion of the digital native debate, see Benini and Murray (this volume); see also Lotherington and Ronda (this volume).
[3] The generation of students born from 1982-2000 are described as Millennials (Howe & Strauss, 2000) and the Net Generation (Oblinger & Oblinger, 2005; Rosen, Carrier & Cheever, 2010; Tapscott, 2009). The younger siblings of the Net Generation have been defined as the iGeneration by Rosen, Carrier, and Cheever.

2012) that provides an overview of the digital native debate and offers a number of useful recommendations:

> As contemporary research begins to expound upon key digital native claims, there is clearly an opportunity for new research that informs theory and practice by investigating whether and how undergraduate learners see value in emerging technologies within their own diverse learning contexts. To this end, perhaps . . . researchers will endeavour to further investigate what so-called Net generation students see as important by asking them about their technology needs and values directly. (p. 14)

Following Smith's recommendation to examine students' practices, beliefs, and preferences (see also Roberts, 2004), this chapter aims to provide administrators and faculty (i.e., language instructors and teacher educators) with a better understanding of the diversity of students' technology-related experiences and skills. In so doing, this chapter will contribute to the ability of administrators and faculty to make informed decisions about the future of the curriculum, faculty and student training, and other related issues.

The following questions guide the present study:

1. To what extent do patterns of technology use of undergraduate language learners align with what it means to be a so-called digital native?
2. How do undergraduate learners use new technologies for academic versus nonacademic purposes?
3. What are some of the new literacy practices that undergraduate students seem to prefer or dislike?

The next part of this chapter offers an overview of selected recent survey-driven studies of new literacy practices and the use of digital tools in various contexts. The results from our surveys at a large public university in the US are then presented in Sections 3 and 4. The chapter concludes with a discussion of the findings and directions for future research.

2. Background

Several studies have examined the extent to which students' uses of new technologies reflect the characteristics of so-called digital natives. Contrary to Prensky's (2001) claim that digital natives constitute a unique generation that is inherently equipped to handle tasks with technology in the classroom, Corrin, Bennett, and Lockyer (2013) discovered a wide range of technological expertise and digital literacy awareness and practices.

> Limited transfer was found between the technologies students used in their everyday life and those they used for their academic study ... [and that] methods in the classroom should not be designed for a single generation with mutual characteristics, but must cater for a variety of different learners. (p. 136)

In an analysis of five different age groups ranging from age 14 to age 65, Helsper and Eynon (2010) observed that age as well as gender, level of education, internet experience, and breadth of internet use predicted digital nativeness. Jones, Ramanau, Cross, and Healing (2010) reported that digital natives are not a homogeneous generation of technology users. Younger students consider the internet to be a more important tool for downloading or streaming video, and they use social networks more frequently than students between the ages of 21 and 25 (p. 727). Margaryan, Littlejohn, and Vojt (2011) demonstrated that digital natives primarily use older (i.e., 1.0) internet technologies (as opposed to web 2.0 technologies) and that the extent of students' technology use and digital literacy practices are influenced to some degree by academic major, faculty integration of technology, and institutional encouragement and support for digital technologies in learning and teaching. Other studies have also highlighted the need for a great deal of additional research on factors that have an impact—either positive or negative—on digital natives' use of digital tools and development of digital literacies (Correa, 2010; Hargittai, 2010; Jones & Healing, 2010; Kennedy, Judd, Dalgarno, & Waycott, 2008; Livingstone & Helsper, 2007, inter alia).

Pasfield-Neofitou (2013) explored the constructs *native speaker* and *digital native* in a study of Australian learners of Japanese (as a foreign language) during real-time, computer-mediated communication (CMC) with Japanese native speakers outside the classroom setting. Results reveal that typing proficiency and the ability to manage more than one thread in the same conversation are more important factors than age for effectively participating in real-time chat with native speakers.

In an extensive review of previous research, Hubbard (2013) provided evidence contradicting the assumption that digital natives do not require training to effectively use technologies for language learning. Moreover, Hubbard (2013) observed that "beyond the perceived value of learner training for making current [computer-assisted language learning] tasks and activities more effective, it also provides a foundation for the development of greater learner autonomy and support for lifelong learning" (p. 174).

Recent CALL research has produced a limited number of survey-driven studies that offer a variety of different types of information about digital tools and digital literacies in foreign and second language education. For this summary of previous research, we have selected six studies that are directly related to CALL, digital literacies, and other dimensions of foreign/second language education. The studies reviewed here were not all conducted in the US, which means that certain factors may be slightly or extremely different from those that affect the ways in which our students perceive and access digital tools. One such example is internet access, which tends to vary greatly at many different levels from place to place. While we acknowledge a range of contexts in the studies reviewed here, we will leave it to readers to evaluate generalizability and the extent to which specific aspects of any particular study may or may not be relevant or transferable.

2.1 Winke and Goertler (2008)

In a survey of technology access and use completed by more than 900 first- and second-year learners of French, German, and Spanish at a US university, Winke and Goertler (2008) discovered that most of the participants could easily send e-mail (93%), navigate the internet (92%), and play a video on a website (87%). However, less than 40% of these students stated that they could easily edit and upload video, make an audio recording, or develop and maintain a website. Although they frequently used social networking sites (85%) and sent text messages (80%) in their personal lives, few students (4%) in the study believed that these tools would be useful for language learning. Only 27% indicated that instructors used public websites in their language classrooms, but 87% of the students used them for personal purposes.

According to Winke and Goertler (2008), language teachers should understand how students currently use online communication spaces and tools outside of the classroom for their personal communication so that they can, in turn, create real-world tasks using many of these same tools for language learning. "Regularly surveying the students will help teachers and administrators design appropriate tasks, harness new technologies students already use in their personal lives, and generate motivation for learning online" (p. 497).

2.2 Winke, Goertler, and Amuzie (2010)

Winke, Goertler, and Amuzie (2010) included the participants from Winke and Goertler (2008) in an expanded study of students enrolled in courses of commonly taught languages (CTLs) ($N = 1,413$) and students enrolled in less commonly taught languages (LCTLs) ($N = 736$). Overall, the results for both CTLs and LCTLs were similar to those reported in Winke and Goertler's 2008 study. More than 90% of all students could easily navigate the internet and send email messages. Fewer than 50% of students in CTLs and LCTLs indicated that they could easily make audio recordings, and fewer than 25% of all participants stated that they could easily develop or maintain a website. Students enrolled in LCTL courses reported lower levels of computer literacy than those learners taking CTL. Winke et al.'s analysis of LCTL learners' comments led them to conclude that the lack of technology use in LCTL classrooms may account—to some extent—for learners' lower ratings of their computer literacy.

2.3 Kim, Rueckert, Kim, and Seo (2013)

In another study conducted at a U.S. university, Kim, Rueckert, Kim, and Seo (2013) compared perceptions and practices of two groups of mobile technology users in a graduate TESOL course in which they completed class activities using mobile technologies: 25 students used mobile devices (e.g., mobile telephones and tablets), and 28 students used mobile computers (e.g., laptops and netbooks). Overall, in the prestudy survey, students in both groups agreed that the use of mobile devices and computers increases access to resources (88%) and that such use also improves communication with instructors and classmates (80%). Results from the prestudy survey also revealed that students in the mobile device group were more open to using new technologies than those in the mobile computer group. Results from a poststudy survey suggested that students in the mobile computer group became more willing to using new technologies for language learning.

Kim et al. (2013) suggested that language educators "should embrace students' perceptions and recognize them as essential when designing effective mobile learning environments" (p. 61). Such findings also align with those of Winke and Goertler (2008) in that both studies emphasize the need to survey language students' preferences for using technologies when instructors create technology-based lessons and tasks. Kim et al. also found that technological affordances and limitations should also be considered when designing tasks using mobile telephones and tablets. For example, some of the students in the postsurvey study commented that they switched from their mobile devices to their mobile computers for completing activities in which they had to read or type comments on their peers' activities.

2.4 Lu, Throssell, and Jiang (2013)

In a survey-driven study of 347 English language learners in mainland China majoring in humanities, sciences, technology/engineering, and fine arts, Lu, Throssell, and Jiang (2013) used a 24-item questionnaire to gain a better understanding of students' preferences for using CALL technologies in and outside the classroom. Results indicated that, from the point of view of the learners, frequent use of CALL was only useful for learning English-language songs outside the formal learning setting. No statistically significant differences were found for academic major or gender. In their analysis of semistructured interviews with 20 of the participants, Lu et al. found 113 comments that expressed negative opinions of working with CALL. In particular, these students stated that they lacked technological skills and that they had difficulty accessing the internet outside of class. The authors recommend that both students and their teachers receive training to use CALL technologies.

2.5 Moroz (2013)

In a 2013 study conducted by Moroz, 139 students enrolled in first- through fourth-year Japanese courses at a Canadian university completed a 23-item survey about their overall awareness of smartphone applications (apps), with a focus on participants' perceptions of the usefulness of two apps in particular. Neither the number of years of language study nor time spent in Japan significantly influenced students' awareness or use of apps for language learning. Dictionaries were the most popular app followed by kanji/kana apps and vocabulary and culture apps.

Students reported that they discovered new apps by exploring an app store (81%), browsing the internet (49%), and by talking with classmates (48%) and their teachers (23%). In the second part of the study, four beginning-level and nine intermediate-level students rated the usefulness of features of a flashcard app (KanjiBox) and a dictionary app (Kotoba!). Overall, students found both apps useful, but beginning-level students found the flashcard app more useful, and intermediate learners preferred the features of the dictionary app. According to Moroz (2013), app developers should elicit feedback from students and teachers; likewise, teachers should become more aware of apps for language learning so that they can share their knowledge of effective apps with their students.

2.6 Steel and Levy (2013)

In another survey-driven study, Steel and Levy (2013) analyzed 587 students' use of 20 technologies inside and outside of language classes specifically for supporting their (foreign) language study. The participants—students at a university in Australia—also ranked which tools were beneficial for language study outside of the classroom. Steel and Levy divided their results into three categories: (a) those tools used by more than 50% of students, (b) those used by 30-49% of the sample and (c) those used by less than 30% of students. For both inside and outside of class use, they discovered that online dictionaries and web-based translators were used the most (more than 80% for both tools), and more than 40% of the students ranked these tools as beneficial for language learning. YouTube (69%), social networking sites (57%), and mobile phone applications (56%) were commonly used for language study outside of class. Tools used by 30% to 49% of the students for studying languages outside of the classroom included Skype, discussion forums, instant messaging, podcasts, blogs, and wikis. Learners generally ranked these tools as less beneficial for language learning than dictionaries and web-based translators. The students in this study were predominantly first- and second-year undergraduate students enrolled in beginning- and intermediate-level courses.

According to Steel and Levy (2013), these learners might have felt less confident about using their limited knowledge of the foreign languages to communicate online. They also suggest that students likely used dictionaries, mobile applications, and social networking sites more often for language study (outside of class) than other tools because these technologies do not require extensive training in order for students to learn how to use them. Less than 30% of students used iTunes U, chat rooms, microblogging sites, or virtual worlds.

2.7 Other Research

A number of other recent studies have focused on students' use and perceptions of a specific tool, for example: iTunes U (Rosell-Aguilar, 2013), tablets (Chen, 2013), mobile devices (Hsu, 2013), and wikis (Wang, 2014). Another line of survey-driven research has examined teachers' use of new technologies and their integration into the foreign language curriculum (Celik, 2013; Lomicka Anderson & Williams, 2011).

2.8 Summary

Although it might be convenient, for whatever reason, to identify individuals as members of a specific, single category (e.g., digital native), it seems clear that human beings are all multifaceted, and we all necessarily have different experiences because we perceive, act, and react in ways that are unique within a range of contexts. The research reviewed in this section confirms that age is certainly much less important as a factor than so many other variables that influence perceptions and practices. Pasfield-Neofitou's (2013) observation about CMC can be extended to the overarching theme of digital literacies:

> CMC competence appears more subtle and complex than simply being a native speaker or a digital native. It requires not only linguistic and technical competence, but also a sensitivity to know how to combine these competences in a variety of situations with a number of different interlocutors. (p. 156)

3. Survey A

This section presents the results of our first survey (Survey A); the results of our second survey (Survey B) are presented in Section 4. We administered both surveys in 2013, and although the results of each study are presented separately, the Discussion section offers a combined set of reflections on the findings.

3.1 Design

The first group of items on Survey A elicits demographic information, which is reported for the primary purpose of allowing future replication, in whole or in part, of the present study (see the survey instrument in Appendix A). Age was the only type of demographic data used to determine the sample population since we had made the decision to focus on language learners between the ages of 18 and 23, the typical age range of university students.

The next two sections (i.e., Technology Proficiency and Use of Online Workbook) use items with a Likert response format including five options (e.g., *Much lower*, *Lower*, *Average*, etc. and *Never*, *Once a week*, *2-3 times per week*, etc.). The third section includes a two-part item numbered as 14a and 14b in order to emphasize the connected nature of these two components and required a response totaling 100% (e.g., a response of 70% for one part would automatically produce a response of 30% for the other part). The results from the final section are not reported here, but they are being used in the second phase of our long-term project.

3.2 Data Collection and Participants

This survey was administered during the Spring 2013 semester to over 800 undergraduate learners of French, German, Italian, Japanese, and Spanish; however, the results from German and Japanese courses were not used for the present study since students in these courses did not use an online workbook—an important part of this survey. Table 1 shows the age distribution of all respondents (i.e., learners of French, Italian, and Spanish) who took the survey.

Table 1
Age Distribution of All Respondents (Survey A, Item 3)

Age	N	%	%
18	22	2.9	
19	127	16.7	
20	162	21.3	84.5
21	178	23.5	
22	89	11.7	
23	64	8.4	
24	27	3.6	
25	30	4.0	
26	12	1.6	
27	14	1.9	14.5
28	8	1.0	
29	8	1.0	
30	3	0.4	
31+	15	2.0	
Total	759	100.0	100.0

During an initial review of all submitted surveys, the authors excluded any that had not been filled out accurately or completely. Next, surveys submitted by anyone under the age of 18 or over the age of 23 (shaded rows in Table 1) were discarded, which means that 84.5% of all respondents represent the sample population of the present study (Table 2). Other demographic information is provided in Table 3 (gender), Table 4 (first language), and Table 5 (academic level).

Table 2
Age Distribution of the Participants (Survey A, Item 3)

Age	N	%
18	22	3.4
19	127	19.8
20	162	25.2
21	178	27.7
22	89	13.9
23	64	10.0
Total	642	100.0

Table 3
Gender Distribution of the Participants (Survey A, Item 2)

Gender	N	%
Female	414	64.5
Male	228	35.5
Total	642	100.0

Table 4
Native Language of the Participants (Survey A, Item 4)

Language	N	%
English	593	92.4
Spanish	26	4.0
French	9	1.4
German	0	0.0
Italian	0	0.0
Other	2	0.3
Total	642	100.0

Table 5
Academic Level of the Participants (Survey A, Item 5)

Academic Level	N	%
Freshman (Year 1)	95	14.8
Sophomore (Year 2)	209	32.6
Junior (Year 3)	235	36.6
Senior (Year 4)	85	13.2
Super Senior (Year 5)	18	2.8
Total	642	100.0

3.3 Survey A Results

Since many new educational (and noneducational) platforms, software packages, and tools are being designed for mobile use, Survey A focuses on mobile devices with internet connectivity. We decided to group smartphones and tablets together in Item 7 to have an overall sense of the extent to which students are advancing toward what has been called "ubiquitous computing" (Poslad, 2009). Although the results in Table 6 indicate that 48 language learners (7.5%) do not own a mobile device with internet connectivity, it is currently possible for them to borrow a laptop from the university's main library.

Table 6 (Survey A, Item 7)
Do you own a mobile device with 3G/4G/wi-fi capability such as a smartphone or a tablet?

Response	N	%
Yes	594	92.5
No	48	7.5
Total	642	100.0

Nonetheless, these students would not enjoy the same amount of access as students who own their own device, so having almost 50 students from the sample population without all-day access would certainly be less than ideal if digital tools for mobile devices were adopted on a wide scale.

The two items in Table 7 shows that the respondents clearly perceive themselves to be much more technologically capable and digitally savvy than the generation of their parents..

Table 7
Technological Proficiency with Desktop/Laptop Computers (Survey A, Items 8 & 9)
Item 8: *Generally, what is your skill level as a user of desktop/laptop computers compared to **typical undergraduate students**?*
Item 9: *Generally, what is your skill level as a user of desktop/laptop computers compared to **people around the age of 50**?*

	Item 8 N (%)	Item 9 N (%)
Much Higher	62 (9.6)	455 (70.9)
Somewhat Higher	242 (37.7)	160 (24.9)
Average	324 (50.5)	20 (3.1)
Somewhat Lower	14 (2.2)	7 (1.1)
Much Lower	0 (0.0)	0 (0.0)
Total	642 (100.0)	642 (100.0)

Over 70% of the respondents chose *Much Higher* for Item 9. However, it is important to recognize that these young adults probably have a wide range of reasons for feeling and answering this way, which is why Phase 2 of our long-term project will explore in more details learners' beliefs, practices, and preferences for this item and others that provide little or no information from a *yes/no* response format

The results for Items 10 and 11 of the survey in Table 8 mirror rather closely the results for Items 8 and 9 presented in Table 7.

Table 8
Technological Proficiency with Mobile Devices (Survey A, Items 10 & 11)
Item 10: *Generally, what is your skill level as a user of mobile devices such as smartphones or tablets with 3G/4G/wi-fi capability compared to **typical undergraduate students**?*
Item 11: *Generally, what is your skill level as a user of mobile devices such as smartphones or tablets with 3G/4G/wi-fi capability compared to **people around the age of 50**?*

	Item 10 N (%)	Item 11 N (%)
Much Higher	57 (8.9)	424 (66.0)
Somewhat Higher	179 (27.9)	166 (25.9)
Average	362 (56.4)	43 (6.7)
Somewhat Lower	29 (4.5)	7 (1.1)
Much Lower	15 (2.3)	2 (0.3)
Total	642 (100.0)	642 (100.0)

For desktop and laptop computers as well as internet-capable mobile devices, our participants between the ages of 18 and 25 feel comparatively more or much more capable than people who are part of their parents' generation (i.e., around the age of 50). However, for Items 8 and 10, the majority of participants perceive that their technological ability with computers and mobile devices is average for their peer group.

The results for Items 12 and 13 (see Table 9) are related to the results presented for Item 14 (see Table 10). The former is an indication of frequency, and the latter shows which tool learners prefer.

Table 9
Frequency of Online Workbook[4] Use (Survey A, Items 12 & 13)
Item 12: *How often do you use a **desktop/laptop computer** to access My French/Italian/Spanish Lab?*
Item 13: *How often do you use a **mobile device with 3G/4G/wi-fi such as a smartphone or tablet** to access My French/Italian/Spanish Lab?*

	Item 12 N (%)	Item 13 N (%)
Every day	5 (0.8)	574 (89.4)
4-5 times per week	58 (9.0)	31 (4.8)
2-3 times per week	361 (56.2)	17 (2.7)
Once a week	150 (23.4)	11 (1.7)
Never	68 (10.6)	9 (1.4)
Total	642 (100.0)	642 (100.0)

[4] The online workbook (MyFrenchLab, MyItalianLab, MySpanishLab) is part of the publisher's textbook bundle.

The results presented in Table 10 are from two related items that required a combined answer totaling 100%. For example, if a student indicated that he/she does 12% of My French/Italian/Spanish Lab assignments with a mobile device (Item 14b), the same respondent was required to answer 88% for Item 14a (i.e., the percentage of assignments completed using a desktop/laptop computer).

Table 10
Accessing the Online Workbook (Survey A, Items 14a and 14b)
Item 14a: *What percentage of your assignments in My French/Italian/Spanish Lab do you do with a **desktop/laptop computer**?*
Item 14b: *What percentage of your assignments in My French/Italian/Spanish Lab do you do with a **mobile device with 3G/4G/wi-fi such as a smartphone or tablet**?*

Percentage of Assignments	Online Workbook Access: Desktop/Laptop Computer N (%)	Online Workbook Access: Mobile Device N (%)
100	596 (92.8)	0 (0.0)
90-99	24 (3.8)	0 (0.0)
80-89	5 (0.8)	1 (0.2)
70-79	8 (1.2)	0 (0.0)
60-69	2 (0.3)	2 (0.3)
50-59	3 (0.5)	0 (0.0)
40-49	0 (0.0)	3 (0.5)
39-39	2 (0.3)	2 (0.3)
20-29	0 (0.0)	8 (1.2)
10-19	1 (0.2)	5 (0.8)
1-9	0 (0.0)	24 (3.8)
0	0 (0.0)	596 (92.8)
Total	642 (100.0)	642 (100.0)

Since 92.8% of students reported doing 100% of their online workbook assignments with a desktop or laptop computer, it is clear that this particular sample of the population might not be ready for any kind of large-scale implementation of a mobile-assisted language learning (MALL) program. Once again, it is important to note that the responses provided by our participants do not explain the reasons why they use one technological tool versus another. In the case of this particular online workbook, the publisher may or may not have created a smartphone or tablet application that would give students the same options or at least options that they would prefer.

4. Survey B

Survey B (the second survey used for the present study) includes items that elicit definitions and explanations of terms related to life in a digital era and the use of specific digital tools (see survey instrument in Appendix B). As noted above, Survey A and Survey B are part of a long-term project during which interviews in the form of focus groups are currently using these survey results as discussion topics in order to explore issues related to digital literacies and the use of digital tools in more detail.

4.1 Design

The first three items in Survey B elicited demographic information, and the only factor used to determine inclusion in the sample was age. The first language of participants was also included as an item in order to determine whether or not our sample population included a disproportionate number of participants whose L1 was not English. Gender was also part of the demographic data collected from this section of the survey for the same reason (i.e., to insure that this dimension of the demographic composition of the sample was not disproportionate in some obvious way).

The second section of the survey (Appendix B) required the most potential effort on the part of the participants since this section included two closed questions (4a, 5a) that were each paired with an open question (4b, 5b). Although the survey was designed to be brief (i.e., taking approximately five minutes for typical respondents to complete), the open questions were placed at the beginning of the survey as one way to increase the potential for substantive responses that might include more than a few words.[5]

In the middle of the survey (the second page), items related to digital device ownership offered *Yes* and *No* as the only options. Given that many libraries and corporations provide students and employees with access to some of these tools, ownership and other types of access will be included in future phases of this study. Two of the items in the Digital Skill Level section were adapted from a survey developed by the Pew Internet and American Life Project (Purcell et al., 2012) that asked teachers to rate their students' literacy practices when using search engines for research.

4.2 Data Collection and Participants

Since students' participation in this project was dependent on each instructor's ability to find time to administer the survey during class, a convenience sample of data including mainly students at the third-year level was collected for the first round of survey distribution in Summer 2013. A second round of survey distribution was planned in advance for the first week of the Fall 2013 semester, which allowed us to reach hundreds of additional learners who were more likely within our target age range since this second round focused on first-semester students.

Surveys completed by anyone under the age of 18 were automatically excluded

---

[5] Some research has shown that placing an open-ended question at the beginning and the end of a survey can produce fuller, nonrepetitive data (Johnson, Sieveking, & Clanton, 1974; see also Herzog & Bachman, 1981); however, this same research recommends placing important open-ended questions at the beginning only in cases where repeating the same open-ended question(s) might unnecessarily lengthen the time required for participation.

(shaded in gray in Table 11) since the possible inclusion of data from minors had not been addressed in the Institutional Review Board application for this project. We also decided to exclude surveys completed by anyone born before 1990 (also shaded in gray in Table 11) in order to have a typical age range of undergraduate students (i.e., those who spend 4-5 years working toward an undergraduate degree). Consequently, the sample population used for the present study includes 608 respondents who reported their age as being anywhere between 18 years, 0 months and 23 years, 6 months (see Table 12 below). These 608 respondents represent 81.2% of all respondents, as shown in Table 11.

Table 11
Age Distribution of All Respondents (Survey B, Item 2)

Age	N	%
< 18	9	1.2
18-23	608	81.2
24-29	112	15.0
30s	15	2.0
40s	4	0.5
50s	1	0.1
Total	749	100.0

Table 12
Age Distribution of the Participants (Survey B, Item 2)

Age	N	%
18	93	15.3
19	149	24.5
20	126	20.7
21	103	16.9
22	87	14.3
23	50	8.2
Total	608	100.0

The gender distribution of respondents is displayed in Table 13. There is not an equal percentage of female and male students; however, the present study does not rely on gender as an independent variable.

Table 13
Gender Distribution of the Participants (Survey B, Item 1)

Gender	N	%
Female	374	61.5
Male	234	38.5
Total	608	100.0

Table 14 provides the distribution of the native languages of the participants.

Table 14
Native Languages of Participants (Survey B, Item 3)

Language	N	%
English	597	98.2
Spanish	3	0.5
English and Spanish	3	0.5
Other	5	0.8
Total	608	100.0

4.3 Survey B Results

We decided to elicit an answer for a question using the term *digital native* simply to see if these young adults themselves would be aware of a term commonly used to characterize them. It is difficult to know if it should be surprising or unsurprising that almost one third of the participants did not identify themselves as digital natives (see Table 15). This may be considered yet another piece of evidence that even if some researchers see these young adults as a monolithic group who all behave in similar ways, the members of this demographic set see themselves more as multifaceted, complex individuals who belong to and participate in a variety of different communities and contexts.

Table 15
Are You a Digital Native? (Survey B, Item 4a)

Response	N	%
Yes	442	72.7
No	166	27.3
Total	608	100.0

Since this item (Item 4a) required participants to make a simple *yes/no* decision, Item 4b (i.e., open response) was included in the survey to allow for clarifications, examples, definitions, and so forth. The themes reported in Table 16 for Item 4b were not predetermined or indicated anywhere on the survey instrument. These categories were created and revised over time by the authors as they transcribed, organized, and reviewed the comments from participants.

Table 16
Define the term *digital native*. (Survey B, Item 4b)

Dominant theme of definition	N	%
Someone "born"/"raised"/"grew up" surrounded by technology in a digital age/environment	290	47.7
Someone who has computer and/or technology proficiency and/or uses computers/technology frequently	182	29.9

Someone who has natural inclinations/intuitions/abilities when using new technology	14	2.3
Someone who learns languages on line or uses their native language to communicate electronically	13	2.1
Someone who prefers spending time and interacting in virtual spaces rather than physical spaces	8	1.3
Someone who understands how to interact in social media	5	0.9
[Miscellaneous]	16	2.6
[No definition provided by respondent. Includes responses such as "No idea." and "I don't know."]	80	13.2
Total	608	100.0%

Although the wording in some of these categories in Table 16 may overlap with some of the other categories, the decision to create specific categories was guided by the recurring use of specific terms. For example, since the phrase "Someone who understands …" was specifically used by five different participants, it was given its own category even if it could have possibly been combined with one of the more popular categories.

The comments below are excerpts from Item 4b: "Define the term *digital native*". The first group includes remarks that characterize digital nativeness as a result of having been born and raised in a period of seemingly ubiquitous digital technologies.

Group A, S40
You have grown up using and around digital technology.

Group A, S42
Growing up using electronic devices.

Group A, S46
Born into technology.

Group A, S56
Someone who has been around technology their whole life and can pick up a laptop, smartphone, tablet and know how to use it.

Group B, S68
Being born in the era of computers.

Group C, S30
Introduced to technology at [a] young age.

Group E, S2
Someone who has been raised around modern technology, such as iPads, Mac computers, Android tablets & phones, etc. A digital native must be "technologically savvy" and must be able to quickly adapt to new & upcoming technology.

Group F, S1
Someone who has grown up with evolving technology.

Group F, S8 A digital native is a[n] individual who's born during the time technology was exposed to the public & is pretty familiar w[ith] the concept [of] it today.

Group G, S123
I have never heard this term before so I can only assume it means educated with technology from a young age.

Group G, S124
A person native to the digital world: who learned to be digitally proficient at the time they learned to speak.

Group G, S125
Someone born in the technology age who can generally understand how to work and navigate technology.

Group G, S134
The ability to navigate the internet, cell phones, and other digital technology because you grew up with it.

Group G, S165
Someone who has used technology their whole life.

Group H, S52
A person who is a part of the generation that is tech savvy.

Although the next group of comments referred to some type of natural inclination regarding the ability to use new technologies, this set of remarks did not emphasize the (potential) importance of having been born into/around new technologies or having grown up with them.

Group E, S48
Someone with common sense to turn on a Mac, smart phone, and not have to use carrier pigeons or use telegraph.

Group F, S26
I have no idea. I'm assuming a person who was born with the ability to comprehend technology.

Group G, S122
Someone who is naturally inclined to use technology.

Group G, S146
One whom digital aspects of life come naturally or easy to acquire the needed knowledge.

Group G, S158
Someone who naturally is good with electronics & likes to use them in everyday life often.

Group H, S51
Naturally good with working with technology.

Other comments focused primarily on the interrelated themes of familiarity and proficiency without necessarily providing an explanation of the ability to use and understand technology.

Group A, S35
Someone who can communicate through technology.

Group A, S1
Somebody who understands technology.

Group A, S6
Someone who knows their way around the internet and is decently well functioning with technology.

Group A, S80
Someone who knows about the latest tech.

Group A, S5
Good with electronics.

Group A, S98
A person who is at home with technology and able to understand it intuitively.

Group B, S78
I do not know what it is. Maybe a technically knowledgeable person.

Group C, S20
Technology literate.

Group H, S11
Familiar with technology.

Group F, S15
Being able to use technology?

Group H, S13
A person who is a "digital native" would be someone who is familiar with the technology that's been recently developed over the last 6 years, at least. They understand computers, cameras, cell phones, smart phones, how to sync such things to other technological devices.

Another set of comments placed an emphasis on liking or preferring new technologies as a reason for being perceived as a digital native.

Group A, S38
Someone who finds a home for themselves more so on the internet, than anything else.

Group B, S94
Technologically inclined?

Group D, S46
A person with many digital devices & high knowledge of all the technology today!

Group E, S49
Someone who loves/needs to have a digital device on them at all times.

Group F, S12
Person who likes digital things.

A rather smaller group of comments was identified as being the result of an understandable misinterpretation of this item. These students simply confused or conflated terms such as *native* and *native speaker*.

Group B, S91
Operates technology in your native language.

Group E, S7
Ability to learn another language on line.

The comments provided below illustrate the types of remarks that were categorized as miscellaneous.

Group A, S34
I don't know what a digital native is. My best guess is that it's someone from another country that communicates digitally to their home country.

G Group, S174
Someone born in the Matrix.

After the two items related to digital nativeness, the survey moved to a pair of items related to digital literacy (see Tables 17 & 18).

Table 17
Please indicate your level of digital literacy. (Survey B, Item 5a)

Level	N	%
Very Low	34	5.6
Low	37	6.1
Medium	187	30.8
High	247	40.6
Very High	103	16.9
Total	608	100.0

Since Item 5a (Table 17) required respondents to select a specific level of digital literacy, Item 5b (Table 18) was created to provide respondents with an opportunity to clarify or explain their own understanding of digital literacy. The categories in Table 18 were created and revised over time by the authors as they transcribed, organized, and reviewed the comments from participants.

Table 18
Define the term *digital literacy*. (Survey B, Item 5b)

Dominant theme of definition	N	%
Ability to understand and use computers/digital devices	478	78.6%
Ability to understand digital language/terms	25	4.1%
Ability to read digital content	24	3.9%
Language learning/development through digital devices	8	1.3%
Learning through digital devices	4	0.7%
[Miscellaneous]	7	1.2%
[No definition provided by respondent. Includes responses such as "No idea." and "I don't know."]	62	10.2%
Total	608	100.0%

The excerpts provided below demonstrate how participants phrased their responses to this open-ended question in Item 5b (see Table 18). The first group of comments has in common the notion that digital literacy involves understanding new technologies and being able to use them as a resource for various purposes. This group represents, by far, the largest category and seems to be a result of two factors. First, students may already be much more familiar with the terms *digital* and *literacy* individually, so it is easy for them to imagine what the two terms together represent. Second, it is possible that students realized that these key words used together represent something general, and they therefore defined and described digital literacy in a general way as they completed the survey. In any case, it is clear that—at least on some level—there was more convergence by participants toward a single definition of digital literacy than was the case for the definition of digital nativeness.

Group A, S3
Understanding how to use online resources efficiently.

Group A, S7
Being able to do basic procedures on a computer or phone & not need the help of your children/grandchildren.

Group A, S10
Digital literacy is how familiar you are with technology.

Group A, S12
Digital slang & uses of digital language to communicate with those that understand.

Group A, S17
The ability to navigate the internet and level of difficulty doing so.

Group A, S18
Using technology as a resource for school (books, homework) and other important tasks that require reading.

Group A, S33
The ability of an individual to communicate, learn, and absorb information through computers, cell phones, and mainly the internet. For example, if you are digitally literate, you'd prefer to research/learn via online sources rather than books from a library.

Group A, S100
Being able to operate, utilize, and navigate through technology.

Group B, S3
How well you understand computers & other electronics.

Group B, S73
Proficiency with modern technology.

Group C, S7
This term seems to be synonymous with computer literacy.

Group C, S13
Digital literacy is a person's understanding of and ability to use concepts, programs, etc. digitally, such as on a computer or tablet.

Group C, S40
Navigating websites & solving online errors.

Group C, S93
Understanding of the digital world/media.

Group D, S10
Capability of using technology and resources available through technology.

Group D, S17
Able to function on a computer.

Group E, S11
[Being] aware of the digital frontier and your ability to interact with the fast growing digital world. In a sense it is how up to date you are with being able to use technology as far as computer or interface devices.

Group E, S33
How competent one is with digital technology.

Group F, S14
Digital literacy covers a wide range of topics, including things like social media, an ability to understand, navigate, and implement software features, dealing with different operating systems & digital devices, and even a component of programming.

Group F, S39
Able to use electronics efficiently.

Group H, S2
The ability to locate, read, and understand digital information.

Group H, S7
Ability to understand and navigate the use of digital technology.

Group H, S13
Being able to utilize technology to the fullest capacity it was meant to be and to be able to share that information in a straightforward way.

Group H, S24
Knowledge of what electronic devices are capable of and what they are used for.

Group H, S31
Able to navigate and understand the digital realm.

Group H, S44
Knowing how to write computer script (java, python, etc.) or being able to read into or modify computer programs.

Group H, S45
Being literate in terms of digital things.

Other comments framed digital literacy as a skill set involving knowledge of specific terms and/or language associated with digital tools.

Group A, S9
Your ability to know terms when using technology.

Group A, S30
Someone who can understand terminology on the internet or with technology.

Group C, S2
The ability to understand and comprehend terms and language that are exclusive to the "digital community."

Group C, S11
Understanding terms used in a digital environment.

Group C, S25
The ability to read technology terms and make sense of it or be familiar with it.

Group D, S25
Digital literacy is understanding the terms used to define aspects of technology.

Group D, S72
Dealing with terms and slang that may involve anything digital.

52 SURVEY OF USE OF DIGITAL TOOLS FOR LANGUAGE LEARNING

Group E, S12
Knowing terms that relate to programs & hardware.

Group E, S18
Digital literacy is the language of technology and how well someone has an understanding of [this language].

Group E, S27
[This is related to] whether or not I understand the common terminology of modern technology.

A much smaller group of comments focused on a more basic definition of digital literacy as something that involves being able to read, and Student 8 (Group E) seems to indicate that he/she is unaware of the accepted academic/technical definition even though the goal of this item was to elicit students' perceptions of what is entailed in digital literacy.

Group D, S68
Like being literate labels a person as being able to read, digital literacy is the ability to understand digital media and technology. I could be very wrong.

Group E, S8
Digital literacy is the ability to read subjects displayed electronically.

Group G, S142
Media reading.

Some other comments related to language learning or learning in general in a digital environment.

Group A, S27
When you learn the basics of a language through technology.
[This student had explained digital native as "When you learn how to speak a language fluently through technology?"]

Group E, S7
Ability to learn on line.

The results for Items 6a through 6f (see Table 19) suggest that a laptop computer and a smartphone with full e-mail and web browsing capability are overwhelmingly the most popular tools among these students. Given that a relatively high percentage of respondents own a laptop and the fact that 32.2% of them own a desktop, the results indicate that most or all of the students who own a desktop computer also own a laptop.

Table 19
Ownership of Digital Devices ($N = 608$) (Survey B, Items 6a-6f)

Device	Yes	No	Total
a) Desktop computer	196 (32.2%)	412 (67.8%)	608 (100.0%)
b) Laptop computer	589 (96.9%)	19 (3.1%)	608 (100.0%)

c)	Cell phone **without** full e-mail and web browsing capability	69 (11.3%)	539 (88.7%)	608 (100.0%)
d)	Smartphone **with** full e-mail and web browsing capability	564 (92.8%)	44 (7.2%)	608 (100.0%)
e)	Tablet that is **not** simply an e-reader	208 (34.2%)	400 (65.8%)	608 (100.0%)
f)	E-reader	112 (18.4%)	496 (81.6%)	608 (100.0%)

The sum of each row in Table 20 is equal to the sample population ($N = 608$). Two comment are in order concerning the organization of Table 20. First, the items have been rearranged in order to display the results from easiest to most difficult, according to the responses. Second, the column *Not Applicable* represents the response selected by anyone who had never used a particular tool/skill listed in the first column of Table 20.

Table 20
Digital Skill Level (Survey B, Items 7a-7l)

Digital Skill Set		Digital Skill Level		
		Easy	Difficult	Not Applicable
j)	Create/send/receive mobile phone text messages	595 (97.9%)	13 (2.1%)	0 (0.0%)
g)	Use a search engine	594 (97.7%)	14 (2.3%)	0 (0.0%)
k)	Download a mobile phone app that I want/need	584 (96.1%)	22 (3.6%)	2 (0.3%)
d)	Use the main features of Facebook	583 (95.9%)	23 (3.8%)	2 (0.3%)
b)	Create a multimedia presentation	564 (92.8%)	42 (6.9%)	2 (0.3%)
h)	Play online single-player games	547 (90.0%)	50 (8.2%)	11 (1.8%)
f)	Use the main features of Instagram or Tumblr (or some other photo-blogging tool)	527 (86.7%)	57 (9.4%)	24 (3.9%)
c)	Upload a video to YouTube (or another video-sharing site)	518 (85.2%)	66 (10.9%)	24 (3.9%)
e)	Use the main features of Twitter	488 (80.3%)	81 (13.3%)	39 (6.4%)

i)	Play online multiplayer games	474 (78.0%)	102 (16.8%)	32 (5.3%)
a)	Typing non-English accents/characters	276 (45.4%)	302 (49.7%)	30 (4.9%)
l)	Use computer coding, programming, and/or programming techniques to create software or modify existing software	115 (18.9%)	442 (72.7%)	51 (8.4%)

The relatively high percentages of *Easy* responses suggest that this sample population has a rather robust confidence level with almost all of these digital skills with the exceptions of typing non-English accents and using computer coding. Nonetheless, by selecting either *Easy* or *Difficult* for Item 7l (computer coding), 91.6% of these language learners have some experience with coding. The fact that they perceive it as difficult is certainly not as much of a barrier for potential use in the foreign language curriculum as it would have been if they had selected *Not Applicable*.

In Table 21, the first column represents the response, from 1 (*very difficult*) to 10 (*very easy*), that was selected by the student in response to each of three questions. The table only provides percentages since they are often easier to compare when viewing results from two or more items that are presented together.

Table 21
Items 8a, 8b, 8c ($N = 608$); Scale of 1 (*very difficult*) to 10 (*very easy*)
Item 8a: Indicate how difficult or easy it is to determine if information that you find on the internet is accurate/reliable.
Item 8b: Indicate how difficult or easy it is to identify the original source of information that you find on the internet.
Item 8c: Indicate how difficult or easy it is to determine the point of view/bias of the information that you find on the internet.

Response	% 8a	% 8b	% 8c
1	0.0	0.3	0.0
2	0.2	1.3	0.0
3	0.8	3.1	0.7
4	2.5	5.8	2.5
5	8.1	10.7	6.6
6	8.9	11.7	11.8
7	21.4	20.7	19.1
8	25.0	19.6	21.9
9	14.8	11.8	18.4
10	18.4	15.0	19.1

One way to interpret this ten-point scale in Table 21 is to group together responses in the *difficult/very difficult* (Response 1-4) section, the middle/neutral section (Responses 5-6), and the *easy/very easy* (Responses 7-10) section of the range of possible ratings. For 8a and 8c, the *easy/very easy* section represents almost 80% of responses (79.6% and 78.5%, respectively). Although not as high, the *easy/very easy* section for Item 8b almost reached 70% of responses. Nonetheless, this means that there are still 20%-30% of respondents who find certain literacy-related abilities to be somewhat or very difficult. For all three items (8a, 8b, 8c), the students seem to recognize certain limitations related to the use of search engines; however, it is possible that they would have more confidence if the items were more closely related to nonacademic topics or projects. Incidentally, as noted above, this will be the goal of the subsequent phases of our long-term project.

In Table 22, it is important to note that a rank of 1 was given by each participant for the resource used *most frequently*. The Frequency Rank column was added to indicate clearly that "Internet resources not mentioned above" has a mean rank that represents resources most frequently used, and "Libraries (going to a library building)" received a mean rank that represents resources least frequently used.

Table 22
Kendall's Coefficient of Concordance: Ranking Frequency of Using Resources for Academic Purposes (Survey B, Item 9)

Type/Location	Mean Rank	Frequency Rank
Internet resources not mentioned above	1.89	1
Wikipedia	2.73	3
Online resources of the university's libraries	2.44	2
Libraries (going to a library building)	2.93	4

This coefficient of concordance is a nonparametric statistical procedure that tests for significance, and in this case, significance was found ($p < .01$), which means that the rankings do not appear to have been randomly assigned by participants. The W statistic[6] (.124) indicates that, although the rankings were not seemingly assigned in a random way by respondents, there was relatively little exact alignment of rankings for all four resources.

Given the rise in popularity of Facebook, Twitter, Instagram, and other social media platforms, tools, and environments, Item 10 (see Table 23) may indeed be one of the most important questions related to short-term curricular planning in today's language programs (and perhaps even in upper level undergraduate courses). These results suggest that the students' opinions on this topic are almost evenly split, but as is the case for most items included in Survey A and this survey,

---

[6] This can be interpreted like a correlation from 0 to 1; a negative correlation is never reported for a typical Kendall's coefficient of concordance procedure.

further inquiry is needed since the forced-choice *yes/no* response format can only provide a general perspective of attitudes, beliefs, and perceptions.

Table 23
Should social media be integrated into the foreign language curriculum? (Survey B, Item 10)

Response	N	%
Yes	318	56.0
No	240	43.0
Total	558	100.0

Given that members of the Net Generation (and soon, the members of the iGeneration) seem to need some type of guidance, we decided to see if students really felt that technology-related training would be important as a tool for language learning. It is important to note that especially for Item 10 (and Item 11 below), it would be extremely beneficial to understand why students chose one option or the other. It would also be helpful to faculty and administrators to know how much such training would cost and who would pay for it (and for how long), among other things.

Table 24
How important is it for you to receive more technology-related training for learning a foreign language? (Survey B, Item 11)

Response	N	%
Not Important	155	27.8
Somewhat Important	323	57.9
Very Important	80	14.3
Total	558	100.0

5. Discussion

The results from both surveys reveal a few clear patterns of digital tool use and preferences. In Survey A, for example, only 48 out of 642 respondents reported not having a mobile device with 3G, 4G, or wi-fi capability.[7] Such a high rate of access to networked technologies suggests that—at least regarding the technological dimension—it would already be possible to integrate one or more types of mobile-assisted language learning assignments or projects to the curriculum. Another key finding is that over 93% of the students who completed Survey A self-reported an average or higher skill level using mobile devices when comparing themselves to their peers. By selecting average or higher, these language learners seem to be indicating, once again, that it is not the technologies themselves that would create any substantial barrier to a foreign language curriculum that in-

---

[7] See Zickuhr and Rainie (2014) for a nationwide survey of e-reader and smartphone ownership and literacy practices.

cluded some type of assignments or projects involving the use of mobile devices. Nonetheless, when asked specifically about how they access the online workbook (i.e., the main digital component currently integrated into the curriculum), almost 93% of these students indicated that they never access the online workbook with a mobile device, and 3.8% reported using a mobile device for only 1%-9% of online workbook assignments. Only about 3% of the respondents indicated that they use a mobile device to do 10% or more of online workbook assignments. This item had been added to the survey because this is one of the issues that needs to be explored throughout all three phases of the long-term project. It will be important to determine why so few students use mobile devices for something that is already an integral part of the curriculum. In other words, if they are not able or willing to use mobile devices for something as basic as an online workbook, this potentially represents some sort of red flag for future assignments or projects that rely on mobile technologies.

The results of Survey B reveal that these language learners did not unanimously agreed that they themselves are considered digital natives, perhaps because this is a term proposed by others and used mainly to create clear-cut categories for the purpose of research. Although many respondents provided definitions of the term *digital native* that closely resemble definitions proposed by demographers and other researchers, the students' definitions expressed a range of ideas that might be associated with the notion of digital nativeness. However, these students had a much more unified view of the definition of the term *digital literacy*, most likely because literacy is something that they have heard, read, seen, and experienced during their (approximately) 15 years of formal education. One of the main findings related to the potential for the future use of mobile-assisted language learning assignments and projects in Survey B is found in the results of Item 6d. For this item, 92.8% of students self-reported having a smartphone with full e-mail and web browsing capability. This rate of smartphone ownership, according to Smith (2013), is much higher than the national average (in the US), which was reported as being at 56% for adults age 18 and older. Smith also indicates that for people between the ages of 18 and 24, ownership of smartphones is at 79% (i.e., much lower than the percentage reported by our participants). Nonetheless, it is important to remember that even if the rate of smartphone ownership within a sample population is high, this does not indicate the type of platform (e.g., iPhone, Android, etc.), and one of the major factors to consider when determining whether or not, and how, to integrate digital tools into the curriculum is the ability to have software that can work on any platform (or any version of that platform). Fortunately, it seems that as time passes, hardware and software developers—and publishers—are becoming more aware of the need for options, even if this results in higher costs or production delays.

Item 10 on Survey B is perhaps one of the most relevant issues for current and short-term planning for language programs. The results for Item 10 (*Should social media be integrated into the foreign language curriculum?*)[8] show that students

---

[8] For a nationwide survey of social media participation, see Duggan and Smith (2013).

are almost evenly split for and against the use of social media in foreign language courses.[9] However, only a few years ago, Winke and Goertler (2008) included similar items on a survey (e.g., Facebook/MySpace; Second Life), and only 4% of their participants indicated that Facebook/MySpace "is/would be useful for language learning" (p. 491), even though 85% reported using Facebook/MySpace for their "personal life" (p. 491). Moreover, only 5% of these same students considered Second Life to be a useful tool for learning a foreign language. This does indeed show that over time, our students' attitudes may continue to move in the direction of seeing social media as pervasive, not limited only to personal use. Even if students already seem to be technologically savvy, one reason for their hesitation to accept social media (and digital tools in general) as a part of the curriculum might be due to the real or perceived burden of having to learn how to use the tools that they already know in a different way. In other words, when anyone uses Twitter, Facebook, Instagram, or any other type of social media, it is up to the individual user to decide how much time to spend using it and which features are the most or least helpful and/or interesting. The challenge for faculty and administrators will be to begin any curriculum revision by determining how digital tools—and which ones—can be used specifically for language learning instead of first deciding to use specific hardware, software, or social media environments and then trying to transform them into tools for language learning and teaching without a sound pedagogical underpinning.

6. Conclusion

As time passes and additional research demonstrates that age is only one of many factors that can influence how and why people use different types of technology in a range of contexts, it may be possible to move beyond the debate surrounding how to characterize and categorize users of digital tools. Smith (2012) offers a nice summary of the types of questions that we should be asking as we seek to understand perceptions and uses of new technologies:

> Much of the criticism regarding the digital native debate underscores a lack of research that authentically maps not only the rapidly shifting technology developments, but also the emergent nature of the perceptions and viewpoints informing the learner, educator, and researcher assumptions and beliefs underlying such debates. Questions remain regarding how we might reframe and reconsider new typologies or constructs around student technology uses, values, and needs, including the following:
>
> - What is the role of the language in both informing and reflecting our perceptions of and experiences with emerging technologies in education …?

---

[9] Winke and Goertler (2008) also asked their participants about specific social media tools/environments. It seems reasonable to compare their results and ours since many of today's social media tools/environments (e.g., Instagram, Twitter) did not exist (or were not as widely used) at the time.

- If there is a new teaching and learning ecology, ... how can we authentically understanding and engage with this ecology beyond the binaries of digital native/immigrant?
- Rather than simply considering technology usage and digital emergences, how might we further understand the various perceptions, values, and perspectives informing discursive debates regarding learning and technology across generations? (p. 14)

References

Celik, S. (2013). Internet-assisted technologies for English language teaching in Turkish universities. *Computer Assisted Language Learning, 26,* 468–483.

Chen, X. B. (2013). Tablets for informal language learning: Student usage and attitudes. *Language Learning & Technology, 17,* 20–36. Retrieved from http://llt.msu.edu/issues/february2013/chenxb.pdf

Correa, T. (2010). The participation divide among "online experts": Experience, skills and psychological factors as predictors of college students' Web content creation. *Journal of Computer-Mediated Communication, 16,* 71-92.

Corrin, L., Bennett, S., & Lockyer, L. (2013). Digital natives: Exploring the diversity of young people's experience with technology. In R. Huang, Kinshuk, & J. M. Spector (Eds.). *Reshaping learning: Frontiers of learning technologies in global context* (pp. 113–138). New York, NY: Springer.

Duggan, M., & Smith, A. (2013). *Social media update 2013.* Washington, DC: Pew Research Center. Retrieved from http://www.pewinternet.org/files/2013/12/PIP_Social-Networking-2013.pdf

Hargittai, E. (2010). Digital na(t)ives? Variation in Internet skills and uses among members of the "Net Generation." *Sociological Inquiry, 80,* 92–113.

Helsper, E. J., & Eynon, R. (2010). Digital natives: Where is the evidence? *British Educational Research Journal, 36,* 503–520.

Herzog, A. R., & Bachman, J. G. (1981). Effects of questionnaire length on response quality. *Public Opinion Quarterly, 45,* 549–559.

Howe, N., & Strauss, W. (2000). *Millennials rising: The next great generation.* New York, NY: Vintage Books.

Hsu, L. (2013). English as a foreign language learners' perception of mobile assisted language learning: A cross-national study. *Computer Assisted Language Learning, 26,* 197–213.

Hubbard, P. (2013). Making a case for learner training in technology enhanced language learning environments. *CALICO Journal, 30,* 163–178.

Johnson, W. R., Sieveking, N. A., & Clanton, E. S. (1974). Effects of alternative positioning of open-ended questions in multiple-choice questionnaires. *Journal of Applied Psychology, 59,* 776–778.

Jones, C., Ramanau, R., Cross, S., & Healing, G. (2010). Net generation or digital natives: Is there a distinct new generation entering university? *Computers & Education, 54,* 722–732.

Jones, C., & Healing, G. (2010). Net generation students: Agency and choice and the new technologies. *Journal of Computer Assisted Learning, 26,* 344–356.

Junco, R. (2012). Too much face and not enough books: The relationship between multiple indices of Facebook use and academic performance. *Computers in Human Behavior, 28,* 187–198.

Kennedy, G., Judd, T., Dalgarno, B., & Waycott, J. (2010). Beyond natives and immigrants: Exploring types of net generation students. *Journal of Computer Assisted Learning, 26,* 333–343.

Kim, D., Rueckert, D., Kim, D.-J., & Seo, D. (2013). Students' perceptions and experiences of mobile learning. *Language Learning & Technology, 17,* 52–73. Retrieved from http://llt.msu.edu/issues/october2013/kimetal.pdf

Lankshear, C., Knobel, M., & Curran, C. (2013). Conceptualizing and researching "New Literacies." In C. A. Chapelle (Ed.), *The encyclopedia of applied linguistics* (pp. 863–870). Hoboken, NJ: Wiley Blackwell.

Leu, D. J., Kinzer, C. K., Coiro, J., Castek, J., & Henry, L. A. (2013). New literacies and the new literacies of online reading comprehension: A dual level theory. In N. Unrau & D. Alvermann (Eds.), *Theoretical models and process of reading* (6th ed.) (pp. 1150–1181). Newark, DE: International Reading Association.

Livingstone, S., & Helsper, E. (2007). Gradations in digital inclusion: Children, young people and the digital divide. *New Media & Society, 9,* 671–696.

Lomicka Anderson, L., & Williams, L. (2011). The use of new technologies in the French curriculum: A national survey of teachers of French. *French Review, 84,* 764–781.

Lu, J., Throssell, P., & Jiang, H. (2013). Exploring the application of computer-assisted English learning in a Chinese mainland context: Based on students' attitudes and behaviours. *International Journal of English Linguistics, 3*(3), 31–41.

Margaryan, A., Littlejohn, A., & Vojt, G. (2011). Are digital natives a myth or reality? University students' use of digital technologies. *Computers & Education, 56,* 429–440.

Moroz, A. (2013). *App assisted language learning: How students perceive Japanese smartphone apps* (Unpublished master's thesis). University of Alberta, Alberta, Canada. Retrieved from http://hdl.handle.net/10402/era.30241

Oblinger, D., & Oblinger, J. L. (Eds.) (2005). *Educating the net generation.* Boulder, CO: EDUCAUSE. Retrieved from http://www.educause.edu/research-and-publications/books/educating-net-generation

Palfrey, J., & Gasser, U. (2008). *Born digital: Understanding the first generation of digital natives.* New York, NY: Basic Books.

Pasfield-Neofitou, S. (2013). Digital natives and native speakers: Competence in computer-mediated communication. In F. Sharifian & M. Jamarani (Eds.), *Language and intercultural communication in the new era* (pp. 138–159). New York, NY: Routledge.

Poslad, S. (2009). *Ubiquitous computing: Smart device, environment, and interactions.* West Sussex, England: Wiley.

Prensky, M. (2001). Digital natives, digital immigrants. Part 1. *On the Horizon, 9*(5), 1–6. Retrieved from http://www.marcprensky.com/writing/Prensky%20-%20Digital%20Natives,%20Digital%20Immigrants%20-%20Part1.pdf

Purcell, K., Rainie, L., Heaps, A., Buchanan, J. Friedrich, L., Jacklin, A., ... Zickuhr, K. (2012). *How teens do research in the digital world.* Retrieved from http://www.pewinternet.org/files/old-media//Files/Reports/2012/PIP_TeacherSurveyReportWithMethodology110112.pdf

Roberts, G. R. (2004). *Technology and learning expectations of the Net Generation.* Washington, DC: EDUCAUSE. Retrieved from http://www.educause.edu/research-and-publications/books/educating-net-generation/technology-and-learning-expectations-net-generation

Rosell-Aguilar, F. (2013). Podcasting for language learning through iTunes U: The learner's view. *Language Learning & Technology, 17,* 74–93. Retrieved from http://llt.msu.edu/issues/october2013/rosellaguilar.pdf

Rosen, L. D., Carrier, L. M., & Cheever, N. A. (2010). *Rewired: Understanding the iGeneration and the way they learn.* New York, NY: Palgrave Macmillan.

Rosen, L. D., Carrier, L. M., & Cheever, N. A. (2013). Facebook and texting made me do it: Media-induced task-switching while studying. *Computer in Human Behavior, 29,* 948–958.

Seely Brown, J. (2002). Learning in the Digital Age. In M. Devlin & J. Meyerson (Eds.), *The internet & the university: Forum 2001* (pp. 65–91). Washington, DC: EDUCAUSE. Retrieved from https://net.educause.edu/ir/library/pdf/FFPIU015.pdf

Smith, E. E. (2012). The digital native debate in higher education: A comparative analysis of recent literature. *Canadian Journal of Learning and Technology, 38*(3), 1–18. Retrieved from http://cjlt.csj.ualberta.ca/index.php/cjlt/article/view/649/347

Smith, A. (2013). *Smartphone ownership 2013*. Washington, DC: Pew Research Center. Retrieved from http://www.pewinternet.org/files/old-media//Files/Reports/2013/PIP_Smartphone_adoption_2013_PDF.pdf

Steel, C. H., & Levy, M. (2013). Language students and their technologies: Charting the evolution 2006-2011. *ReCALL, 25*, 306–320.

Tapscott, D. (2009). *Grown up digital: How the net generation is changing your world*. New York, NY: McGraw-Hill.

Wang, Y. (2014). Using wikis to facilitate interaction and collaboration among EFL learners: A social constructivist approach to language teaching. *System, 42*, 383–390.

Winke, P., & Goertler, S. (2008). Did we forget someone? Students' computer access and literacy for CALL. *CALICO Journal, 25*, 482–509.

Winke, P., Goertler, S., & Amuzie, G. L. (2010). Commonly taught and less commonly taught language learners: Are they equally prepared for CALL and online language learning? *Computer Assisted Language Learning, 23*, 199–219.

Zickuhr, K., & Rainie, L. (2014). E-reading rises as device ownership jumps. Washington, DC: Pew Research Center. Retrieved from http://www.pewinternet.org/2014/01/16/e-reading-rises-as-device-ownership-jumps/

Appendix A
Survey A

NEW TECHNOLOGIES AND LANGUAGE LEARNING

This survey should take you about 5 minutes to complete.
Thank you for providing information that will be used to improve undergraduate education.

BACKGROUND INFORMATION

1. Please provide the name and number of your course: _____

2. Gender (circle one):   FEMALE   MALE

3. What is your current age? _____ YEARS   _____ MONTHS

4. What is your native language (circle one)?
   English   Spanish   French   German   Italian   Other: _____

5. What is your academic level? (circle one)
   Freshman / Sophomore / Junior / Senior / Super Senior (5+ years undergraduate) / Graduate

6. What is (are) your declared major(s)? _____
   If you have not yet declared a major, simply write "N/A" on the line above.

7. Do you own a mobile device with 3G/4G/wi-fi capability such as a smartphone or tablet?
   (circle one)   YES   NO

TECHNOLOGY PROFICIENCY

8. Generally, what is your skill level as a user of **desktop/laptop computers compared to typical undergraduate students**? (circle one)

   Much lower / Somewhat Lower / Average / Somewhat Higher / Much Higher

9. Generally, what is your skill level as a user of **desktop/laptop computers compared to people around the age of 50**? (circle one)

   Much lower / Somewhat Lower / Average / Somewhat Higher / Much Higher

10. Generally, what is your skill level as a user of **mobile devices such as smartphones or tablets with 3G/4G/wi-fi capability compared to typical undergraduate students**? (circle one)

    Much lower / Somewhat Lower / Average / Somewhat Higher / Much Higher

11. Generally, what is your skill level as a user of **mobile devices such as smartphones or tablets with 3G/4G/wi-fi capability compared to people around the age of 50**? circle one)

    Much lower / Somewhat Lower / Average / Somewhat Higher / Much Higher

USE OF ONLINE WORKBOOK (MyFrenchLab / MyItalianLab / MySpanishLab)

12. How often do you use a **desktop/laptop computer** to access MyFrench/Italian/SpanishLab? circle one)

    Never   Once a week   2-3 times per week   4-5 times per week   Every day

13. How often do you use a **mobile device with 3G/4G/wi-fi such as a smartphone or tablet** to access MyFrench/Italian/SpanishLab? (circle one)

    Never   Once a week   2-3 times per week   4-5 times per week   Every day

SURVEY OF USE OF DIGITAL TOOLS FOR LANGUAGE LEARNING

Please make sure that the percentages for 14a and 14b add up to 100%.

14a. What percentage of your assignments in MyFrench/Italian/SpanishLab do you do with a **desktop/laptop computer**? _____

14b. What percentage of your assignments in MyFrench/Italian/SpanishLab do you do with a **mobile device with 3G/4G/wi-fi such as a smartphone or tablet**? _____

Before going any further, please make sure that the percentages for 14a and 14b add up to 100%.

USE OF 3G/4G/WI-FI MOBILE DEVICES

15. What are the **3 most frequent things** (apps/tools) you use your 3G/4G/wi-fi mobile device to do for anything **RELATED TO LANGUAGE LEARNING**?
Please DO NOT include any information related to MyFrench/Italian/SpanishLab.
a. _____
b. _____
c. _____

16. What are the **3 most frequent things** (apps/tools) you use your 3G/4G/wi-fi mobile device to do?
Please **do not** include "talking on the phone" as one of these things.
This question is about **everything**, not language learning specifically.
a. _____
b. _____
c. _____

Appendix B

Survey B

DIGITAL LITERACY AND LANGUAGE LEARNING

This survey should take you 5 minutes to complete.
Thank you for providing information that will be used to improve undergraduate education.

BACKGROUND INFORMATION

1. What is your gender?   FEMALE   MALE

2. What is your current age?   _____ YEARS _____ MONTHS

3. What is your native language?   English   Spanish   Other: _____

4a.   Are you a **digital native**?   Yes   No

4b. Explain in detail in the space below how you would **define a digital native**.

5a. Please indicate your level of **digital literacy** (according to your own understanding of this term).

VERY LOW   LOW   MEDIUM   HIGH   VERY HIGH

5b. To the best of your ability, please provide a definition and explanation of **digital literacy**.

DIGITAL DEVICE OWNERSHIP

6. Please indicate whether or not you own any of the devices listed below.

	Yes (I own this type of device.)	No (I do **not** own this type of device.)
(a) Desktop computer	☐	☐
(b) Laptop computer	☐	☐
(c) Cell phone **without** full e-mail and web features	☐	☐
(d) Smartphone **with** full e-mail and web browsing capability	☐	☐
(e) Tablet that is <u>not</u> simply an e-reader	☐	☐
(f) E-reader	☐	☐

SURVEY OF USE OF DIGITAL TOOLS FOR LANGUAGE LEARNING

DIGITAL SKILL LEVEL

7. Please indicate your ability to do the following. (Select **ONE** response in each row.)
   For any task below that you have <u>never done before</u>, please cross it out.

	This task ranges from somewhat to very **difficult**.	This task ranges from somewhat to very **easy**.
(a) Typing non-English accents/characters	☐	☐
(b) Create a multimedia presentation (with text and video or text and photos).	☐	☐
(c) Upload a video to YouTube (or another video-sharing site).	☐	☐
(d) Use the main features of Facebook.	☐	☐
(e) Use the main features of Twitter.	☐	☐
(f) Use the main features of Instagram or Tumblr (or some other photo-blogging tool).	☐	☐
(g) Use a search engine.	☐	☐
(h) Play online single-player games.	☐	☐
(i) Play online multi-player games.	☐	☐
(j) Create/send/receive mobile phone text messages.	☐	☐
(k) Download a mobile phone app that I want/need.	☐	☐
(l) Use computer coding, programming, and/or programing techniques to create software or modify existing software.	☐	☐

8. On a scale of <u>1 to 10</u>, indicate how difficult or easy it is to …

   a) Determine if information that you find **on the Internet** is accurate/reliable.

   1   2   3   4   5   6   7   8   9   10
   Very Difficult   Very Easy

b) Identify the original source of information that you find **on the Internet**.

1	2	3	4	5	6	7	8	9	10
Very Difficult									Very Easy

c) Determine the point of view/bias of the information that you find **on the Internet**.

1	2	3	4	5	6	7	8	9	10
Very Difficult									Very Easy

DIGITAL RESOURCES & EDUCATION

9. Please provide a ranking for the following items in Column A by matching each item with a ranking from Column B. This ranking is related to the **frequency** with which you use specific resources for any academic purpose.

Column A	Ranking from Column B	Column B (Use each number ONCE.)
Libraries (going to a library building)		1 (Ranking for the item in Column A that you use **most frequently** for research)
Libraries' online resources (databases/books/articles accessible through their website)		2 (Ranking for the item in Column A that you use **second most frequently** for research)
Wikipedia		3 (Ranking for the item in Column A that you use **third most frequently** for research)
Internet resources **not mentioned above**		4 (Ranking for the item in Column A that you use **least frequently** for research)

10. In your opinion, should <u>social media</u> be integrated into the <u>foreign language curriculum</u>?

    YES   NO

11. How important is it for you to receive more technology-related training for learning a foreign language? (This does **NOT** refer to MySpanishLab, MyFrenchLab, or Blackboard.)

    Not important   Somewhat important   Very important

Chapter 3

Challenging Prensky's Characterization of Digital Natives and Digital Immigrants in a Real-World Classroom Setting

SILVIA BENINI AND LIAM MURRAY
University of Limerick (Ireland)

Abstract

In 2001, the terms *digital native* and *digital immigrant* were introduced by Prensky (2001a), and since then, they have been widely used and accepted in various contexts, including education. Prensky has argued that students today, the so-called digital natives, have been immersed in technology all their lives developing technical skills and learning preferences for which traditional education is not well prepared. As such, young people's use of Information and Communication Technologies differentiates them from their teachers, who are digital immigrants. Indeed, the analogy introduced by Prensky to describe today's students and teachers is very appealing; however, no significant empirical evidence exists to support this conjecture, and neither facts nor evidence tested in everyday practice have been provided. This paper aims to reflect on the current uses and expectations of Information and Communication Technologies in education environments and to explore the current debate surrounding Prensky's theory. Consequently, this chapter seeks to provide a critical perspective on the digital natives/digital immigrants divide presenting some of the findings from a major case study in secondary level environments. By monitoring and interviewing students and teachers of two targeted schools, it is intended to examine the actual attitudes and uses of digital technology and digital information and to present recommendations informing best practice for teachers, learners, and second-level institutions.[1]

1. Introduction

The 21st century challenges us with new choices, new perspectives, and opportunities due to the ubiquitous presence of technology in many areas of our lives. The Information Age—or Digital Age—in which we are living has allowed rap-

---

[1] The term 'second-level education' refers to the second-level education sector in Ireland that comprises secondary, vocational, community, and comprehensive schools consisting specifically of a 2-year junior cycle followed by a 2- or 3-year senior cycle.

id global communications and networking to shape modern society; important changes in information access along with parallel technological developments has opened up and redefined the core concept of the delivery of learning in educational institutions. Students' expectations have changed, and strong transformations are taking place (Gibson, 2001). Traditional teaching and learning paradigms and approaches have been, in fact, shaken by the integration of Information and Communication Technologies (ICT) into educational practices.

During the past 20 years, the use of ICT in education has had a rapid development placing some pressure on schools to renovate their approaches and benefit from ICT's educational potential. ICT expectations are very high both at policy and institutional levels. At a policy level its importance has been stressed in sustaining competitiveness in the global economy and market; at an institutional level it is seen as a potential catalyst for a change in education (Ottesen, 2006). Many studies on the use of technology in education consistently found that students in technology rich environments experienced positive effects on performance in all subjects' areas (Lau & Sim, 2008). It has been pointed out that ICT provides fast and accurate feedback to students and speed up computations and graphing, thus freeing the students to focus on strategies and interpretations. Furthermore, the use of interactive multimedia software, for example, can motivate students and lead to improved performance (Jones, 2004). Research also indicates that the use of ICT in education can promote deep learning and allow schools to respond better to the varying needs of students (Barak, 2006). Despite all the apparent benefits of the use of ICT in educational environments, many studies have shown that the learning potential of ICT is not being fully exploited. Many schools, for example, tend to assimilate, rather than accommodate, new approaches to the use of ICT (Higgins & Moseley, 2001). It has been argued also that despite the changes in society as a result of ICT, it is not widely integrated into the educational system and, where it is present and available, there is no evidence that it has affected teaching approaches (Levin & Wadmany, 2005). Hayes (2007) observes that although research into the use of ICT in education is into its third decade there is still "a pressing need to better understand how computer-based technologies are influencing learning opportunities" (p. 385). There are many reasons given for the low level of ICT impact in the classroom; many of the most common factors include inadequate infrastructure, limited access to technology, lack of training and personal expertise, weak technical support, poor planning, and teachers' beliefs (Ringstaff & Kelley, 2002; Baek, Jung, & Kim, 2008). Research categorizes these barriers into two different orders. First-order barriers, the most visible and easiest to remove, include elements such equipment, time, training, and support. Second-order barriers, on the other hand, are more difficult to address since they interfere with or inhibit the process of change; these barriers are rooted in teachers' beliefs about teacher-student roles, classroom practices, teaching methods, and organizational and management methods (Ertmer, 2005). That being said, it is important now to explore the status of ICT in Irish education environments.

2. Background
2.1 ICT Integration in Irish Secondary Schools

The majority of countries have experienced important investments in school ICT facilities, and the Irish government is no exception, having invested significantly in the provision of ICT resources at both primary and secondary level (Mulkeen, 2003). The integration of ICT in postprimary schools was marked by the launch of the Schools IT2000 initiative. The Department of Education and Science, influenced by trends to integrate ICT in teaching and learning globally and concerned about Ireland's economic competiveness in a global information-based society, introduced the initiative in November 1997. The initiative aimed to ensure that all students achieved computer literacy and that teachers were supported in renewing skills which would enable them to integrate ICT in the learning environment. The initiative involved the distribution of IT resources to schools, the provision of IT in-service courses to teachers and the provision of support to schools through the establishment of the National Centre for Technology in Education (NCTE). The support included also the provision of regional IT advisers, the development of internet-based resources and the implementation of pathfinder projects to explore models of best practice (McGarr, 2009). The initiative significantly raised the profile of ICT in education across the country. However, in the following years it appeared to lose a bit of perspective. For example, an evaluation of Schools IT2000 (National Policy Advisory and Development Committee, 2001) reported a substantial increase in IT infrastructure and computer use within schools and found that 59% of the teaching staff availed themselves of the training offered. However, the report also identified the need for a more clearly defined policy in relation to ICT. In December 2001, the Minister for Education announced details of a new 3-year plan which aimed to invest €109 million in ICT in primary and post primary schools. Among other things, some of the aims of the plan included the development of networking infrastructure in all schools, the introduction of broadband access, further development of teachers' skills, and the improvement of multimedia resources for use in schools. In 2005, a national census about the integration of ICT in schools was conducted (Shiel & O'Flaherty, 2006), and the results indicated that ICT resources had increased in schools, but the rate of increase had slowed since 2002. In 2006, the NCTE reported that 96% of schools had been provided with broadband access. Despite the relatively high levels of ICT resources reported, no further policy announcements were made for few years, and a significant decline in ICT investments and general ICT activity in schools took place. In February 2007, the Minister for Education and Science announced the allocation of €252 million for investment in ICT education over a period of 6 years (McGarr, 2009). More recently (2011-2012), a survey dealing with access to, use of, and attitudes toward ICT in education was conducted by the Department of Communications Networks, Content, and Technology of the European Commission Directorate General in 31 European countries. According to this report, students in Ireland benefit from infrastructure and connectivity levels close to the EU mean, and schools with broadband are above the EU average.

Further, ICT use by teachers is considerably above that of other countries, and their overall confidence in using ICT is above the EU mean, except at grade 11. Frequency of students' use of computers is close to EU averages, but their overall confidence levels are below the EU mean (Balanskat, Blamire, & Kefala, 2006; Wastiau et al., 2013).

It seems that the investment in ICT has been generally welcomed by institutions and educators, but this availability of new resources has led this research to investigate how ICT is being integrated into the classroom and to examine ICT uses and expectations for both educators and learners.

2.2 Digital Natives and Digital Immigrants: Theories and Debate

There are words and definitions that we remember particularly well because they seem to easily describe some specific phenomena, finding common ground and aspects of what appears to be an actual reality. *Millennials* (Howe & Strauss, 2000), *Net Generation* (Oblinger & Oblinger, 2005; Tapscott, 1999), *Generation Y* (Jorgensen, 2003; McCrindle, 2006; Weiler, 2005), and finally *digital native* and *digital immigrant* (Prensky, 2001a) are some of the appealing terms with which many of us have become familiar. The idea behind these terms is that a fundamental break has occurred between young people and previous generations and, consequently, between students and their teachers. The authors mentioned above and many others have argued that because today's generation of young people have been immersed in a world infused with technology they behave and learn differently from their predecessors. It is claimed also that they think differently, that they exhibit different social characteristics, and that they have different expectations about life and learning. The new generation of students is said to prefer receiving information quickly, relying on communications technologies for accessing information and interacting with others, favoring active rather than passive learning, being often proficient in multitasking and having low tolerance for lectures (Oblinger & Oblinger, 2005; Tapscott, 1999). It is now important to look specifically at the definitions of digital native and digital immigrant, according to Prensky's ideas and theories as well as the debate that has arisen around this matter in recent years. The terms *digital native* and *digital immigrant* were coined by the US technologist Prensky, appearing first in two seminal articles (2001a, 2001b). Prensky described the generation of young people born since 1980 as digital natives due to a presumed innate confidence in using technologies, having always been surrounded by and continually interacting with them.

> Today's students ... represent the first generation to grow up with new technology. They have spent their entire lives surrounded by and using computers, videogames, digital music players, video cams, cell phones, and all the others toys and tools of the digital age. Our students today are all "native speakers" of the digital language of computers, video games and the Internet. (Prensky 2001a)

Prensky supports his argument by providing some facts and figures, such as students spending fewer than 5,000 hours of their lives reading, but over 10,000 hours playing video games and 20,000 hours watching television. For those born before 1980, Prensky (2001a) coined the term digital immigrants. He claims that this large part of the population, which includes most teachers, lacks the technological fluency of digital natives and, consequently, find themselves unfamiliar with many of the technological skills possessed by students. Digital natives, according to Prensky, process information quickly, multitask, and enjoy gaming, whereas digital immigrants process information slowly, work on one thing at a time, and do not appreciate less serious approaches to learning (this, in practice, means not going to the internet first for information; printing things out as opposed to working on the screen; and reading manuals rather than working things out online). The divide between the interests and technological skills of students and the limited and simplistic use by educators is claimed to be creating alienation and disaffection among students (Levin & Arafeh, 2002; Prensky, 2005). Prensky (2001b) also explains the phenomenon of neuroplasticity, claiming that the brains of digital natives are different from those of previous generations because of the direct effects of digital technologies.

> Based on the latest research in neurobiology, there is no longer any question that stimulation of various kinds actually changes brain structures and affects the way people think, and that these transformations go on throughout life. The brain is ... massively plastic. It can be, and is, constantly reorganized ... . The brain constantly reorganizes itself all our child and adult lives, a phenomenon technically known as neuroplasticity. ("Neuroplasticity," para. 2)

Children raised with computers, as Prensky (2001b) argues, think differently than the rest of us, developing hypertext minds.

Above, all, Prensky underlines his concern about the profound gap he sees between digital native students and their digital immigrant teachers; the natives, according to him, are crying out "for new approaches to education with a better 'fit'" ("What Have We Lost?," para. 4). Young people now have a range of different preferences, tools, and ways of processing and using information that does not fit well with current educational practices. Thus, currently employed pedagogies are outdated and need to be changed. A powerful teaching method, Prensky suggests, would be to use computer games to teach the digital natives. More recently the author has started to move a little bit away from the digital natives-digital immigrants distinction, suggesting instead that digital wisdom should be the organizing feature of our thinking.

> Digital technology, I believe, can be used to make us not just smarter but truly wiser. Digital wisdom is a twofold concept, referring both to wisdom arising *from* the use of digital technology to access cognitive power beyond our innate capacity and to wisdom *in* the prudent use of technology to enhance our capabilities. (Prensky, 2009)

Unlike the digital natives/digital immigrants metaphor, digital wisdom overcomes generational boundaries; even though digital immigrants can never become digital natives, they can acquire and possess digital wisdom. This concept attempts to integrate the immigrants into the technological areas where the natives reside.

All the terms introduced by Prensky instantly became highly popular in public and political debates. The definitions for digital natives and digital immigrants have appeared in numerous articles, blog posts, columns, and books, in general contexts (Bennett, Maton, & Kervin, 2008) and higher education contexts (Jones & Shao, 2011). A quick Google search using these terms provided 238,000 results, and a Google Scholar search, 34,200. While there has been considerable interest in outlining the characteristics of the new generations of students and their learning preferences, there has been little empirical support for several of the claims being made (Jones & Shao, 2011). Many of the arguments about the technological skills, educational preferences, and approaches of the digital natives have been based on conjecture and assumptions (Bennett, Maton, & Kervin, 2008) and lack empirical research. The same can be said about the so-called digital immigrants. These terms became part of our collective knowledge base without having been thoroughly explored in their true nature and everyday practice. Furthermore, very few comparisons of students' and teachers' perceptions of the use of technology have been offered (Waycott, Bennett, Kennedy, Dalgarno, & Gray, 2010), especially in secondary level institutions. A growing body of academic research reveals some of the variables that consitute the stereotypical digital native. Geography, for example, seems to be a very important factor. In the US, there is a different level of web technology and computer use than among the same demographic of digital natives in Australia (Kennedy, Judd, Dalgarno, & Wycott, 2010; Margaryan, Littlejohn, & Vojt, 2011) and in the UK (Stoerger, 2009). In South Africa, only 26% of the population might be described as digital natives (Brown & Czerniewicz, 2010). Research also indicates that socioeconomic factors as well as race, gender, and educational background play an important role in the way in which and the extent to which people use technology (Broos & Roe, 2006). Another important aspect to be considered is age to explain generational differences between digital natives and digital immigrants. Prensky considers this a basic aspect of his theory, while, for other authors, a digital native is defined by exposure to—or experience with—technology (Tapscott, 1999). For some other writers it does not matter much if it is age or experience that defines someone as being a digital native (Oblinger & Oblinger, 2005). Finally, the access to technology and the use of it in both quality and quantity is a significant parameter. While the proportion of young people who use the internet and other new technologies is greater than the older population (Cheong, 2008; Dutton & Helsper, 2007), there are significant differences in how, why, and how effectively young people use these technologies (DiMaggio & Hargittai, 2001; Hargittai & Hinnant, 2008).

This chapter aims to add to this discussion by providing evidence on how students and teachers in secondary level institutions access and use the internet and other new technologies for their language learning and teaching. The data presented will enable us to examine the basic assumptions of the digital natives/digital immigrants, develop a better understanding of the current digital status of our students and teachers, and add to the debate about what and how we should be educating young learners.

3. Method

The research reported here is part of a Ph.D. project. The study employed a mixed-methods approach (Creswell & Clark, 2007) conducting semistructured interviews together with surveys and classroom observations. The case study was conducted in two secondary schools both located in the Munster region, Republic of Ireland. The first school (School A) is a mixed community school (i.e., a type of secondary school funded individually and directly by the state with both male and female students) that accommodates 900 students and a staff of almost 70 people. School A offers progressive educational programs focusing particularly on science, languages, ICT and an overall commitment to innovation and heavy use of ICT in their pedagogies. They even go so far as to use a "flipped classroom" approach in many of their classes (Berrett, 2012; Houston & Lin, 2012) The majority of students are equipped with notebooks or tablet computers as are all of their teachers. The second school (School B) is a Catholic female school with more than 400 students and 40 staff members. In School B, the environment and the teaching reveal a more traditional book-based approach with small class sizes, a close teacher-students relationship, and limited access to a single computer lab for all classes. The participants of the study were second-, third-, and fifth-year students and their Italian and Irish Language teachers. The data elicitation phase lasted 18 weeks and started by asking the participants to complete a preinterview survey to discover their perceptions and uses of ICT inside and outside the classroom. From the 12th to the 15th week, semistructured and focus group interviews were held to investigate the students' and teachers' access and use of ICT focusing particularly on Irish and Italian learning and teaching. During the teachers' interviews the digital natives/digital immigrants divide was discussed in depth. In this paper, we introduce some of the data that arose from both surveys and interviews, presenting selected quotations and relevant analysis in order to address our research concerns and offer future recommendations for teachers, learners, and second-level institutions.

4. Results and Discussion

The main questions presented to the students by both the surveys and the interviews were respectively: *"Do you own any piece of technology?"* and *"Do you have access to any of the following piece of technology: laptop, computer, mobile phone? If yes, how often do you use it/them?"* The responses were unanimous. All students in both schools reported owning a mobile phone and a laptop (shared in some cases) and reported using these tools regularly on a daily basis.

It appears that young people have a great range of ICT available and tend to use technological tools and the internet regularly. In some respects, these first findings seem to support the arguments put forward by Prensky about a large proportion of young people using technologies and the internet as daily "companions." Furthermore, as Prensky has argued, the majority of respondents seem to come from media-rich homes.

Participants were asked to indicate in the surveys how important the use of ICT in their schools was. Students in School A (a total of 41 completed surveys) and School B (a total of 56 completed surveys) students provided very different answers, as figures 1 and 2 indicate.

Figure 1
Students' Perceived Importance of ICT in School A

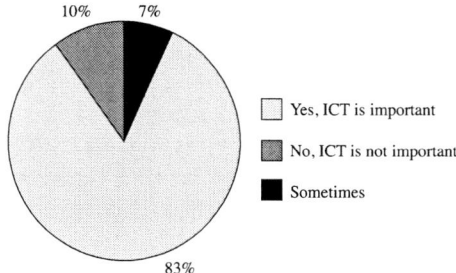

The School A respondents reported the great importance of ICT in their institution. The majority of them (83%) openly addressed the importance of ICT as something being strongly emphasized within the school and widely used for the different subjects.

- *Yes, it is vital for our development in school to see what is going on* (Male 3rd year student, School A).
- *Yes, ICT is very important. It is often used for presentations in class and notes are often put on the school website. Quite a big emphasis is put on ICT and in 4th year, for example, we can complete a Microsoft Office Specialist Course* (Female 5th year student, School A).

Ten per cent of the respondents reported that ICT was not as important as the school suggested but that some teachers relied on it for teaching methods.

- *Not to me specifically but some teachers rely on it for teaching methods* (Female 5th year student, School A)
- *ICT is emphasized but not important* (Male 5th year student, School A)

The rest of students (7%) argued that ICT was not essential in their learning process but that it would definitely help in some cases and specific subjects.

Figure 2
Students' Perceived Importance of ICT in School B

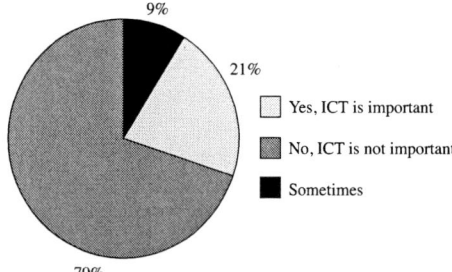

Seventy per cent of the respondents in School B strongly that ICT was not important in their school primarily because the lack of resources and facilities. Students reported the use of books and traditional tools for their learning, but they also felt that ICT should be promoted and used in their learning experience.

- *No. We do not use ICT often so it is not very important, but I think it should be important* (Female 2nd year student, School B)
- *It isn't because there is one computer room for the whole school and we rarely use it* (Female 2nd year student, School B).
- *No, because they don't supply interactive white boards and iPads. We do have computers but we rarely use them. It is not important but it should be* (Female 2nd year student, School B).

Twenty-one per cent of respondents reported that ICT was important in their school for specific subjects. Younger teachers used it for presentations or showing videos.

- *In some classes depending on the subject; some of the young teachers would introduce it in classes showing power points* (Female 3rd year student, School B).

Nine per cent of students stated that ICT was not important in their learning environment except only in certain cases and specific subjects.

When students were asked whether ICT was important for their language acquisition (specifically Italian and Irish), the answers were diverse (see Figures 3 and 4).

Figure 3
Students' Perceived Importance of ICT for Language Learning in School A

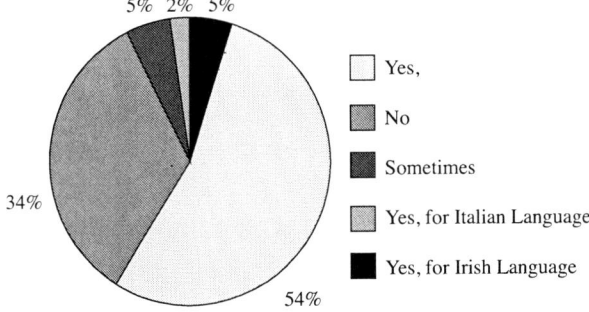

Figure 3 shows that more than half of respondents in School A (54%) stated that ICT is important for their language learning process because it helps to investigate different methods of study by broadening the learning and making it easier. Web tools and the internet are frequently used by the students for their language acquisition. These tools represent for them a great source of information beyond the book and an important support to rely on.

- *Yes, it is important; not all the information are in the book and it is very helpful to have technology to fall back on* (Male 5th year student, School A)

Thirty-four per cent of participants declared that ICT was not particularly important for their language learning mainly because of the lack of available resources.

- *No, not at the moment. There are not many resources available* (Female 5th year student, School A).

Five per cent of the students said that ICT was not essential but could sometimes help the language acquisition process. Another 5% of the students argued that ICT is particularly important for the Irish language especially for downloading information and activities, and 2% of them thought that Moodle and school websites are important and widely used for the Italian Language.

Figure 4
Students' Perceived Importance of ICT for Language Learning in School B

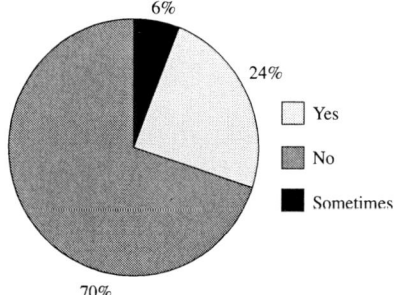

The majority of students in School B (70%) stated that ICT is essential for language acquisition; during the Irish and Italian language classes, digital technologies are not used. Students rely on teachers and books for their learning.

- *Not vital. I think you need to be taught how to speak a language. You can't learn from being on a computer learning about it* (Female 2nd year student, School B).

Twenty-four per cent of students agreed that ICT is important for their language learning. Particularly useful for them are, for example, online dictionaries, videos, and PowerPoint presentations as different tools to approach the language.

Six per cent of participants reported that ICT could be sometimes useful, especially at home as an extra support for the language activities.

- *In school we don't really use ICT resources during language classes. For me, at home, I do use the Internet to help me with languages* (Female 5th year student, School B).

Students in both schools reported that technology is not always essential to learning but they recognize its importance and the fact that some tutors depend on it. Echoing this sentiment, some tutors agreed (but older teachers disagreed).

- *Very important. It is the conduit through which I can reach the students* (Male Irish Language Teacher, School A)
- *I do not make great use of ICT in my classes. I am teaching for over 25 years and tend to fall back to traditional teaching methods. Occasionally I use ICT. Lack of resources is a definite obstacle* (Female Irish Teacher, School B).
- *It is important. However, students do not always enjoy the use of ICT. Some students dislike Powerpoints and prefer if you do not use ICT* (Female Irish Teacher, School B)

The teachers were interviewed on the digital native/digital immigrant divide and were asked if today's teachers and students recognize this concept. Some of their answers reveal deep similarities to those of their students.

- *I'm not one for tags, It's unfair. You are what you are. Digital Native and Immigrants makes no sense, you're not labelled because of your date of birth. There are many people who are engaging technology and they have become literally dependent on it. Many people are just not interested in technology. They* [students] *have not been trained. My own daughter she was born with the computer but that doesn't necessarily mean that you assume all this automatically just because you're born next to a computer. That doesn't necessarily mean that you understand it.* (Female Italian Language Teacher, School B)
- On digital immigrants: *Certainly it is changing because newly qualified teachers are not immigrating towards it* [technology], *they're there really. But certainly we still have the older teachers in teaching struggling, either avoiding it or catching up or learning like myself* (Male Irish Language Teacher, School B).
- On digital natives: *In my opinion it's just a trendy name and I hear this a lot; if you want to put a title on it, well they are digital natives because they are born with it, it is in their environment, they are using the technology but the technology is using them as well; it is not the case for them to be productively using the technology or choosing and controlling the technology that, to me, would be a Digital Native, whereas a lot of people are just passive* (Male Irish Language Teacher, School B).

The findings reported here seem, in some cases, to support the arguments put forward by Prensky and others. Young students appear to be generally enthusiastic about integrating ICT in their learning experience. According to them, ICT helps make the learning more engaging and interactive, and the idea of full ICT integration is generally very appealing for the students of both schools, but they still feel strongly about the importance of a traditional education approach. That is, ICT is important for them but not essential. The traditional book still holds great import and is, together with the teachers, the element students fully rely on. Overall, ICT is seen as a significant "extra support" for learning, an extra tool which learners can easily make use of (e.g., the use online dictionaries for language learning activities at home). Therefore, learners in both schools fit only apparently with the digital natives stereotype. Teachers reported a general enthusiasm about the integration of ICT in their teaching, however they have great concern about the potential technical problems and the appropriate resources available.

This study has depicted a number of benefits provided by the close observation and analysis of the uses of technologies in secondary level environments. Nonetheless, it is necessary to say that this research project has some limitations in terms of generalizations of results. Because of the limited number of schools and participants involved in the project, findings may not be statistically generalizable. This being said, the theories and recommendations presented in this study may be of use for researchers and practitioners.

5. Conclusions and Recommendations

The initial findings of this research highlight the notion that the digital divide students and teachers face is not as simple as Prensky has suggested. In everyday life, all participants use many of the same technologies (e.g., mobile phone, tablets, Web 2.0, etc.), but the types of activities they undertake and the concerns they have are very different. This difference became clear when approaching the educational environment issue. For some students, the idea of using technologies for language acquisition was stimulating, but not essential.

All teachers had a positive attitude toward ICT as a pedagogical device, but many teachers are still not fully trained in the use of iCT and do not make use of it in their teaching. Furthermore, many teachers may exploit ICT for their own learning, but they seem to be very cautious about integrating the use of technology into their own teaching. Overall, teachers recognize the potential of technology to stimulate students' learning and make school studies relevant to real-life situations, but they do not think that ICT is preferable to class-based instruction for promoting cooperation and reflecting on learning, echoing the findings of other researchers (Barak, 2006).

In one school in our study, there remains a very strong traditional book-based and teacher-centred approach; however, this does not reflect a negative attitude towards ICT inclusion. Access to and appropriate training in ICT appear to be the greater concerns in both schools.

The digital divide does not seem to be as clearly defined as Prensky has argued. While there are differences in how generations engage with technologies, there are also similarities across generations mainly based on how much experience people have with the tools. Generational distinctions between natives and immigrants are not reflected in our empirical data, and the uncritical use of these terms could have negative implications for teacher and student interactions. Furthermore, there are significant differences within cohorts of young people in terms of their preferences, skills and use of new technologies (Kennedy et al., 2010) as well as their teachers. As Facer and Furlong (2001) have observed, there is a need

> to critique the concept of the "cyberkid" and the perception of a homogeneous "generation of digital children" in general … . Consequently, we would argue that we need to recognize that childhood is socially and culturally situated, and that different children, like different adults, will have diverse experiences

of and attitudes towards new technologies, experiences that need to be identified and catered for. (p. 467)

In our study, several teachers and students felt that, while technology use was indeed effective, it was not always essential to learning. To our minds, this is a more leveled definition of digital wisdom: the ability to better engage learners and use technology when appropriate, without blindly accepting the latest putative teaching software and while reflecting on its benefits and potential pitfalls.

In order to support future developments in educational contexts, several suggestions can be made:

1. Be aware of *how* and *what* students learn from their teachers.
2. Be aware of the teachers' skills; in some cases secondary school teachers must turn to technicians or librarians for localized support. It is important that teachers continue to develop their own ICT skills and knowledge.
3. Understand what learners are using technologies for, how well technologies are integrated in the educational system, and how familiar teachers are with these tools.
4. Avoid the uncritical use of terms that are not based on empirical evidence because they may have a negative impact upon the perceived possibilities of teacher-student interaction.
5. Be aware that, in many cases, technology is perceived as a very important tool but is not always essential.

Acknowledgments

Strong support for this project has been provided by the School of Languages, Literature, Culture and Communications, University of Limerick. The authors would like to acknowledge the support, availability, and collaboration of the staff members of the two schools in question and all their students.

References

Baek, Y., Jung, J., & Kim, B. (2008). What makes teachers use technology in the classroom? Exploring the factors affecting facilitation of technology with a Korean sample. *Computers & Education, 50,* 224–234.

Balanskat, A., Blamire, R., & Kefala, R. (2006). *The ICT impact report.* Brussels, Belgium: European Schoolnet. Retrieved from http://insight.eun.org/shared/data/pdf/impact_study.pdf

Barak, M. (2006). Instructional principles for fostering learning with ICT: Teachers' perspectives as learners and instructors. *Education and Information Technologies, 11,* 121–135.

Becker, H. S. (1996). The epistemology of qualitative research. In R. Jessor, A. Colby, & R. A. Shweder (Eds.), *Ethnography and human development: Context and meaning in social inquiry* (pp. 53–71). Chicago: University of Chicago Press.

Bennett, S., Maton, K., & Kervin, L. (2008). The "digital natives" debate: A critical review of the evidence. *British Journal of Educational Technology, 39,* 775–786.

Berrett, D. (2012, February 19). How "flipping" the classroom can improve the traditional lecture. *The chronicle of higher education.* Retrieved from http://chronicle.com/article/How-Flipping-the-Classroom/130857/

Broos, A., & Roe, K. (2006). The digital divide in the playstation generation: Self-efficacy, locus of control and ICT adoption among adolescents. *Poetics, 34,* 306–317.

Brown, C., & Czerniewicz, L. (2010). Debunking the "digital native": Beyond digital apartheid, towards digital democracy. *Journal of Computer Assisted Learning, 26,* 357–369.

Cheong, P. H. (2008). The young and techless? Investigating Internet use and problem-solving behaviors of young adults in Singapore. *New Media & Society, 10,* 771–791.

Creswell, J. W., & Clark, V. L. P. (2007). *Designing and conducting mixed methods research.* Thousand Oaks, CA: Sage.

DiMaggio, P., & Hargittai, E. (2001). From the 'digital divide' to 'digital inequality': Studying Internet use as penetration increases. *Princeton University Center for Arts and Cultural Policy Studies, Working Paper Series, 15.* Retrieved from https://www.princeton.edu/~artspol/workpap/WP15-DiMaggio+Hargittai.pdf

Dutton, W., & Helsper, E. (2007). *The Internet in Britain: 2007.* Oxford, England: Oxford Internet Institute, University of Oxford. Retrieved from http://oxis.oii.ox.ac.uk/sites/oxis.oii.ox.ac.uk/files/content/files/publications/oxis2007_report.pdf

Ertmer, P. A. (2005). Teacher pedagogical beliefs: The final frontier in our quest for technology integration? *Educational Technology Research and Development, 53,* 25–39.

Facer, K. and R. Furlong (2001). Beyond the myth of the "cyberkid": Young people at the margins of the information revolution. *Journal of Youth Studies, 4,* 451–469.

Gibson, I. W. (2001). At the intersection of technology and pedagogy: Considering styles of learning and teaching. *Journal of Information Techology for Teacher Education, 10,* 37–61.

Hargittai, E., & Hinnant, A. (2008). Digital inequality differences in young adults' use of the internet. *Communication Research, 35,* 602–621.

Hayes, D. N. (2007). ICT and learning: Lessons from Australian classrooms. *Computers and Education, 49,* 385–395.

Higgins, S., & Moseley, D. (2001). Teachers' thinking about information and communications technology and learning: Beliefs and outcomes. *Teacher Development, 5,* 191–210.

Houston, M. & Lin, L. (2012). Humanizing the classroom by flipping the homework versus lecture equation. In P. Resta (Ed.), *Proceedings of Society for Information Technology & Teacher Education International Conference 2012* (pp. 1177-1182). Chesapeake, VA: AACE.

Howe, N., & Strauss, W. (2000). *Millennials rising: The next great generation.* New York: Vintage.

Jones, A. (2004). *A review of the research literature on barriers to the uptake of ICT by teachers.* London: British Educational Communications and Technology Agency. Retrieved from http://dera.ioe.ac.uk/1603/1/becta_2004_barrierstouptake_litrev.pdf

Jones, C., & Shao, B. (2011). *The net generation and digital natives: Implications for higher education.* York, England: The Higher Education Academy. Retrieved from http://www.heacademy.ac.uk/assets/documents/learningandtech/next-generation-and-digital-natives.pdf

Jorgensen, B. (2003). Baby Boomers, Generation X and Generation Y? Policy implications for defence forces in the modern era. *Foresight, 5*(4), 41–49.

Kennedy, G., Judd, T., Dalgarno, B., & Wycott, J. (2010). Beyond natives and immigrants: Exploring types of net generation students. *Journal of Computer Assisted Learning, 26,* 332–343.

Lau, B., & Sim, C. (2008). Exploring the extent of ICT adoption among secondary school teachers in Malaysia. *International Journal of Computing and ICT Research, 2*(2), 19–36.

Levin, D., & Arafeh, S. (2002). *The digital disconnect: The widening gap between Internet savvy students and their schools.* Washington, DC: Pew Internet & American Life Project. Retrieved from http://www.pewinternet.org/~/media//Files/Reports/2002/PIP_Schools_Internet_Report.pdf.pdf

Levin, T., & Wadmany, R. (2005). Changes in educational beliefs and classroom practices of teachers and students in rich technology-based classrooms. *Technology, Pedagogy and Education, 14,* 281–307.

Margaryan, A., Littlejohn, A., & Vojt, G. (2011). Are digital natives a myth or reality? University students' use of digital technologies. *Computers & Education, 56,* 429–440.

McCrindle, M. (2006). *New generations at work: Attracting, recruiting, retraining & training Generation Y.* Bella Vista, NSW, Australia: McCormick Research. Retrieved from http://mccrindle.com.au/resources/whitepapers/McCrindle-Research_New-Generations-At-Work-attracting-recruiting-retaining-training-generation-y.pdf

McGarr, O. (2009). The development of ICT across the curriculum in Irish schools: A historical perspective. *British Journal of Educational Technology, 40,* 1094–1108.

Mulkeen, A. (2003). What can policy makers do to encourage integration of information and communications technology? Evidence from the Irish school system. *Technology, Pedagogy and Education, 12,* 277–293.

National Policy Advisory and Development Committee. (2001). *The impact of Schools IT2000.* Dublin, Ireland: National Centre for Technology in Education. Retrieved from http://www.ncte.ie/npadc/ncte_report.pdf

Oblinger, D., & Oblinger, J. L. (Eds.). (2005). Educating the net generation. Boulder, CO: EDUCAUSE. Retrieved from https://net.educause.edu/ir/library/pdf/pub7101.pdf

Ottesen, E. (2006). Learning to teach with technology: Authoring practised identities. *Technology, Pedagogy and Education, 15,* 275–290.

Prensky, M. (2001a). Digital natives, digital immigrants. Part 1. *On the Horizon, 9*(5), 1–6.

Prensky, M. (2001b). Digital natives, digital immigrants. Part 2. Do they really think differently? *On the Horizon, 9*(6), 1–6.

Prensky, M. (2005). Engage me or enrage me: What today's learners demand. *EDUCAUSE Review, 40*(5), 60. Retrieved from https://net.educause.edu/ir/library/pdf/erm0553.pdf

Prensky, M. (2009). H. sapiens digital: From digital immigrants and digital natives to digital wisdom. *Innovate: Journal of Online Education, 5*(3). Retrieved from http://www.wisdompage.com/Prensky01.html

Ringstaff, C., & Kelley, L. (2002). *The learning return on our educational technology investment.* San Francisco, CA: WestEd RTEC. Retrieved from http://www.wested.org/online_pubs/learning_return.pdf

Shiel, G., & O'Flaherty, A. (2006). NCTE 2005 census on ICT infrastructure in schools: Statistical report. Dublin, Ireland: National Centre for Technology in Education. Retrieved from http://www.erc.ie/documents/ncte_2005_census_on_ict_infrastructure_ in_schools.pdf

Stoerger, S. (2009). The digital melting pot: Bridging the digital native-immigrant divide. *First Monday, 14*(7). Retrieved from http://firstmonday.org/ojs/index.php/fm/article/ view/2474/2243

Tapscott, D. (1999). Educating the net generation. *Educational Leadership 56*(5), 6–11.

Wastiau, P., Blamire, R., Kerney, C., Quittre, V., Van de Gaer, E., & Monseur, C. (2013). The use of ICT in education: A survey of schools in Europe. *European Journal of Education, 48,* 11–27.

Waycott, J., Bennett, S., Kennedy, G., Dalgarno, B., & Gray, K. (2010). Digital divides? Student and staff perceptions of information and communication technologies. *Computers & Education, 54,* 1202–1211.

Weiler, A. (2005). Information-seeking behavior in Generation Y students: Motivation, critical thinking, and learning theory. *The Journal of Academic Librarianship, 31,* 46–53.

Chapter 4

Agency and Web 2.0 in Language Learning: A Systematic Analysis of Elementary Spanish Learners' Attitudes, Beliefs, and Motivations about the Use of Blogs for the Development of L2 Literacy and Language Ability

JUAN PABLO JIMÉNEZ-CAICEDO
Columbia University (USA)

MARÍA EUGENIA LOZANO
Barnard College (USA)

RICARDO L. GÓMEZ
Universidad de Antioquia (Colombia)

Abstract

This mixed-methods study provides a holistic understanding of literacy practices fostered through an interactive blog project designed for learners in four sections ($N = 54$) of a university-level elementary Spanish course. Data included students' blog posts and an online survey. The analysis drew mainly on Q methodology as a tool for systematically analyzing participants' subjectivities about the use of the blog for developing their Spanish academic literacy and cultural awareness. The findings suggest that students' agency—an individual's socioculturally mediated capacity to act—played an important role in the three different ways students utilized this web 2.0 technology and related electronic literacies in their Spanish course. One group of participants perceived the blog mainly as a very useful tool for practicing their grammar, vocabulary, and new expressions. Another group believed the blog was an important tool for motivating real language use with the support of more expert others. A third group saw it as a space where they could freely interact with an audience and explore or experiment with the foreign language rather than seeing the blog as a place where grammar corrections from instructors were expected. We conclude the chapter with a discussion of practical implications regarding the utilization of blogs for developing second language literacy.

1. Introduction

Over the last few years, web 2.0 applications and their associated digital literacies have been increasingly (re)defining our understanding of teaching and learning languages in classroom contexts at all levels. Such applications have become key and most likely indispensable tools for many language students (McBride, 2009) and teachers. Users are no longer passive recipients of the content; instead, they are often actively engaged in constructing it (Dippold, 2009; Lomicka & Lord, 2009). In addition, web 2.0 tools have given language teaching an emphasis on second language use (rather than second language learning) that covers not only the new advances of web 2.0 technologies, but also the new uses we are giving to existing ones (Johnson, 2004, Warschauer & Grimes, 2007; Tu, Blocher, & Roberts, 2008; Zhang, 2009). Therefore, since web 2.0 technologies represent the most current state of CALL (Walker, Hewer, & Davies, 2008), their role in second language (L2) learning research has been steadily and rapidly growing.

Weblogs (*blogs* hereafter), as one type of web 2.0 application, are asynchronous, interactive, and content-rich environments that allow visitors to post comments to blog authors or other blog visitors. The online environment of blogs enables language instructors to design multimodal, content-based thematic tasks in order to expose students to authentic and appropriate language and cultural materials while allowing all blog members to express themselves in their L2 outside the classroom.

The purpose of this chapter is to analyze in a systematic way university students' attitudes, beliefs, and motivations related to the use of educational blogs for enhancing the content- and task-based language curriculum (Snow, 2001) as well as to discuss the impact this tool may have on students' foreign language learning, communication, and cultural awareness. The research questions addressed in this chapter are the following:

1. How do students view the use of the blog as a tool that can mediate the development of their Spanish (academic) literacy and cultural awareness?
2. How do these students use the literacy-based tasks (reading and writing) of the blog project to expand their knowledge of Spanish language and culture?

2. Background

Previous classroom-based research on using blogs for L2 learning has focused primarily on the impact of technology in the language classroom (e.g., Abdous, Camarena, & Facer 2009; Benito-Ruiz, 2009; Chapelle, 2001). Studies have noted a positive effect for language teaching and learning and, most notably, have documented a shift of dynamics in terms of the affordances the use of technology offers in the classroom, such as increased oral participation influenced by the use of blogs, instructors' pedagogical modifications allowing technology to be a tool rather than a goal, enhanced writing skills, and improved communication skills, among other benefits (Chen, 2009; Deuschemann, Panichi, & Molka-Danielsen, 2009; Dippold, 2009; Egbert, Huff, McNeil, Preuss, & Sellen, 2009; Lee, 2010; Miyazoe & Anderson, 2010; Zhu, 2009).

Furthermore, in a recent review of the literature on web 2.0 and second language learning, Wang and Vásquez (2012) explain that over the last 15 years, there has been a paradigm shift in the language learning/acquisition research, moving from a cognitive orientation to a social one, where naturalistic settings, a participation metaphor, and L2 use rather than L2 learning are the norms. This paradigm shift, they argue, concurs with the main characteristics of web 2.0 technologies. These authors also found that blogs and wikis are the most studied language learning tools to date with second language writing being the most investigated area. For instance, some studies have pointed out that feedback from teachers is the kind of feedback most valued by students, (Dippold, 2009; Hewings & Coffin, 2007; Ware & Warschauer, 2006). This is generally due to the fact that even though students are being immersed in the latest technologies, they continue with the traditional idea of a writing task for which they write a text and then expect to receive feedback from their instructor. Dippold (2009) suggests that in order to fully engage with mediums such as blogs, participants need to abandon traditional roles and conventional writing models.

With regard to the present study, which focuses on the analysis of students' views of the use of blogs in foreign language learning, only a few previous similar studies have been identified. Dippold (2009), for example, looked at students' and teachers' general perceptions of peer feedback through blogs in an advanced German class. She analyzed data produced by nine students in the form of blogs entries, focus group discussions, and their responses to a 5-point Likert-type scale questionnaire with five statements designed to elicit the perceived benefits of and opinions about the use of a blog in the course. The results of her study showed that students enjoyed doing their homework, valued the interactivity of the blog, and were motivated by the medium.

Another study (Miyazoe & Anderson, 2010) analyzed outcomes and students' perceptions of the use of forums, blogs, and wikis within three blended-learning sections of an English as a foreign language course with 61 students at a Japanese university. Drawing on a mixed-methods analysis of a printed questionnaire, interview data, and text analysis, the results indicated a general appreciation for web 2.0 tools as well as overall enjoyment and usefulness of them. In this case, students were most satisfied with wikis, followed by blogs and then forums for all three aspects. Specifically, analysis of lexical density of the students' blog posts "indicated that students' vocabulary became much richer over the course of two semesters in the blogs" (p. 191).

Similarly, Lee (2010) designed a study in which 17 advanced Spanish university students participated in a blog activity as one of their class requirements. Students maintained a personal blog for the duration of the semester (14 weeks) and posted one or two entries each week. They also read, commented on, and responded to each other's entries. Data collected included students' blog postings, online surveys, and interviews. The online survey conducted at the end of the semester included 13 Likert-style statements eliciting information in three areas: the effectiveness of the blogs for online writing and exchange, the affordance of task type, and the role of feedback. Lee's study concluded that the students perceived the use of the blog in a very positive way. Most of the students (80%) agreed that the blog helped them with their Spanish and preferred the blog compared to the regular type of writing assignment. Students also agreed that feedback from the instructor was still very important to them and that their peers' feedback was useful.

While these studies have offered valuable insights for understanding some of the affordances of web 2.0 technologies and students' general perspectives of the use of blogs in (foreign) language classes, no study has yet delved into the systematic exploration of elementary Spanish students' attitudes, beliefs, and motivations regarding the use of blogs for foreign language development. As Wang and Vásquez (2012) have pointed out in their literature review on using web 2.0 technologies in L2 learning, research on the applications of these technologies is still quite limited. Thus, the present study builds on this body of literature and attempts to identify particular aspects of language learning that students believe are impacted the most by the use of a blog in their Spanish course. This study also seeks to understand how learners perceive blog technology as an enhancement or hindrance to learning a foreign language.

3. Theoretical Orientation

3.1 Sociocultural Perspectives

This study is informed by sociocultural perspectives of language and language learning in which literacy is primarily understood as a set of social and cultural practices enacted by a group (Barton, Hamilton, & Ivanič, 2000; Bloome, Katz, Solsken, Willett, & Wilson-Keenan, 2000; Gee, 2000; Lantolf, 2000; Street, 1995). Building on Vygotsky's sociocultural theory of mind—extended by Cole (1996) and Wertsch (1985)—and discussing the application of this theoretical perspective to the study of second language learning, Lantolf states that "the most fundamental concept of sociocultural theory is that the human mind is mediated" (p. 1). Lantolf (1994) further explains that "mediation, whether physical [for example, tools] or symbolic [for example, language], is understood to be the introduction of an auxiliary device into an activity that then links humans to the world of objects or to the world of mental behavior" (p. 418). An illustration of this principle is how we use physical tools such as a pencil and paper as the means to write a text that allow us to materialize our thoughts and ideas. Similarly, a blog in a language course mediates students' learning of the target language through their reading, writing, and interacting in the L2 outside the classroom.

In terms of learning, sociocultural theory maintains that "development does not proceed as the unfolding of inborn capacities, but as the transformation of innate capacities once they intertwine with socioculturally constructed mediational means" (Lantolf & Pavlenko, 1995, p. 109). These sociocultural perspectives have brought attention to the importance of particular human action (e.g., agency) in situated cultural contexts and the way language literacy is intimately bound up with institutions such as schools and social relationships as well as multiple ways of interacting with or around texts (e.g., literacy practices) in L2 classrooms. Such literacy practices usually require students to appropriate new discourses, that is, new ways of talking, thinking, believing, knowing, acting, valuing, feeling, and interacting with multiple texts in various modes of communication as meditating tools; these, in turn, are critical to the development of students' new digital or media literacies (Barton & Lee, 2013; Gee, 1996; Thorne & Reinhardt, 2008).

Thus, from a sociocultural perspective on learning, it is presumed in this study that literacy-based blog interactions (interpreting and producing texts) can promote language learning. As Ryshina-Pankova and Kugele (2013) argue, "learning to perform mental tasks through language use [such as in a course blog] in a dialogue with an expert and peers leads learners to acquire new knowledge" (p. 184). In the context of the present study, the literacy-based and content- and task-based activities designed for the course blog were used as sociocultural tools to mediate students' Spanish literacy development through interaction with peers and more expert others (i.e., instructors and students with greater command of Spanish in the same course) and to increase their cultural understanding while promoting and/or drawing on digital literacies.

3.2 A Literacy-Based Approach to Language Instruction

From a sociocultural perspective on learning, the New London Group (1996) proposed a refinement of literacy education, multiliteracies, which views literacy as a process of negotiating a multiplicity of discourses and plurality of texts that circulate in the context of today's cultural and linguistic diversity. Furthermore, Kern (2000) suggests a literacy or literacies-based approach to language instruction, in which

> [a]cademic language teaching must foster literacy, not only in terms of basic reading and writing skills, but also in terms of a broader discourse competence that involves the ability to interpret and critically evaluate a wide variety of written and spoken texts [in the course blog], physical and digital artifacts of concrete, observable language in use available for interpretation; e.g. video, songs, images, newspapers, books, online discussion sections, gestures, etc. (p. 2).

Indeed, these authors call for a reexamination of the processes and practices of L2 reading and writing—what they call the textual dimensions of digital literacies—as these take place in online environments such as blogs with their new affordances and capabilities while requiring or drawing on students' digital literacies because, as Barton and Lee (2013) argue, most of young people's "digital practices are textually mediated. Producing and using texts online occupies much of their lives" (p. 154; see also Ito et al., 2010).

In the following section, we present the study, starting with a description of its context and the kinds of content- and task-based blog activities in which students participated as part of the literacy-based approach to language instruction that informs our work.

4. The Study

4.1 Context of the Study: An Elementary Spanish II Course and a Multimedia Blog

The Elementary Spanish II course is the second semester of the language program at the university where the present study was conducted. The course curriculum covers topics and communicative functions such as talking about food and restaurants, describing and comparing cities and places, talking about past events and the circumstances surrounding them, discussing professional activities and professions, contrasting present and past learning experiences, and describing people's personalities. A short story and a full-length movie are also part of the course curriculum.

Accordingly, to complement both the course content and specifically the literacy component of the course, the instructors implemented a course blog. The goals of this multimedia blog were to expose students to the language and cultural productions of the Hispanic world and to use these multimodal texts (e.g., videos, newspapers, interactive websites, etc.) as activators of background knowledge in order to encourage students to use the language communicatively while interacting in writing about these texts. Thus, the blog became an integral component of the course by substituting one of the four typical compositions that students were required to produce during the semester. Every other week, students were assigned one thematically driven task in the form of blog prompts. The tasks were designed to have students interact with authentic texts in Spanish and to encourage them to use course-appropriate vocabulary and grammar-related concepts as a way to develop their command of the language and their understanding of cultural topics. These pedagogical tasks required students to do the following:

1. read a text or short article or watch and listen to an embedded video clip;
2. write a description, state their personal opinions, narrate an event or an anecdote, or even share a reflection on the weeks' topic presented in the prompt; and
3. go back to the blog, read a few of their peers' comments and briefly comment, ask, or respond to questions on two or three of their peers' work during the last three days of the blog unit.

4.2 Example of blog task

By the middle of the semester, students had already studied verbal aspect, specifically the different tenses commonly used to express past events in Spanish. Therefore, one of the thematically driven tasks for the blog was created based on the short story *Un día de estos* 'One of these days' by Colombian author Gabriel García Márquez (see Figure 1). After reading the story as homework (and answering vocabulary and comprehension questions), students were presented with a brief lecture about the sociohistorical context of the story, after which they engaged in an in-class discussion about the characters, the plot, and other aspects of the story. Students were then asked to complete the blog task for that particular week.

Figure 1
Sample Screenshot of a Course Blog Page

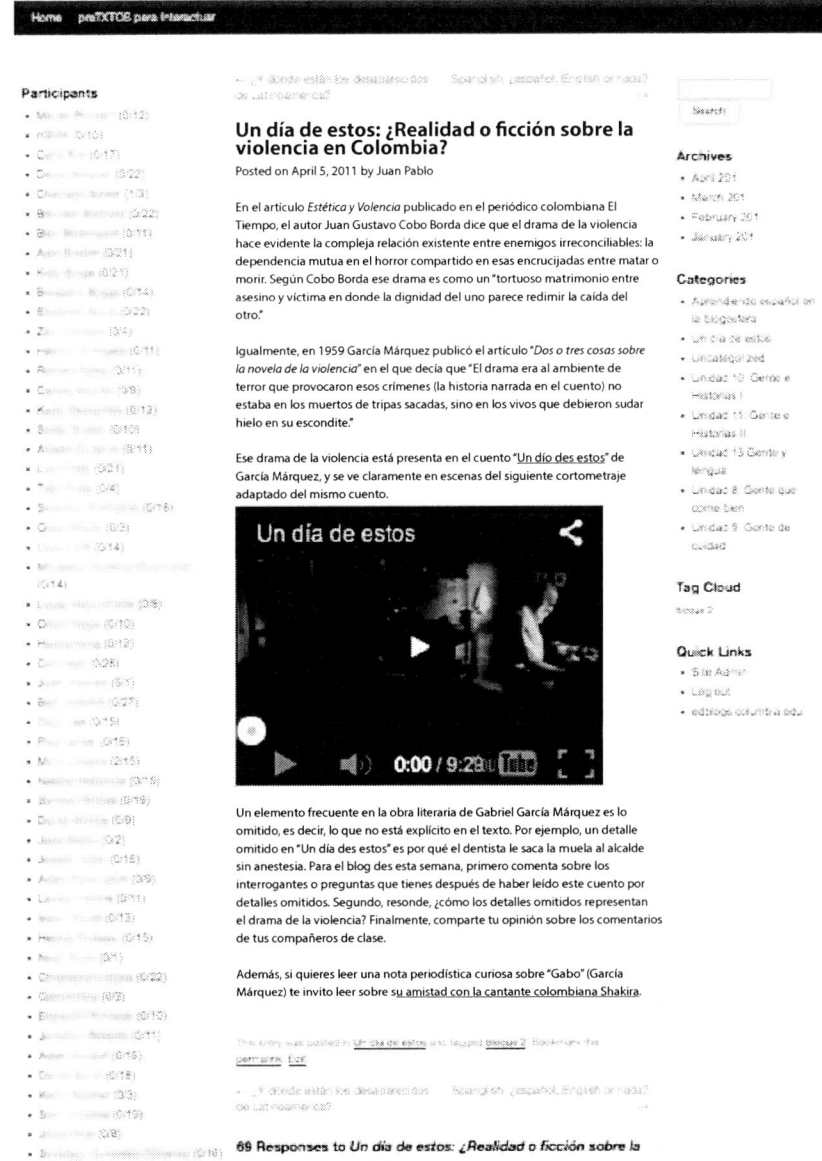

The blog prompt presented a summary of selected newspaper articles about violence in Colombian history, including hyperlinks to the original articles for students who wanted to extend their reading on this topic. The blog prompt also included an embedded YouTube video of a short movie adaptation of *Un día de estos* as a way to expose students to the same text in a different mode (i.e., spoken language) and in a different discourse genre (i.e., a drama). Finally, the blog included a link to an optional news article about the friendship between García Márquez and Colombian pop singer Shakira. This article had the purpose of bringing students back from the images of the violence in Colombian history (depicted in the documentary clips used as part of the lecture and the actual short story) to the present. Therefore, the blog prompt was carefully designed to allow students to interact with authentic texts while being exposed to the grammatical forms and lexicon that they had been studying in the previous and current units.

Regarding the cognitive and linguistic demands of the blog task, students needed to draw on linguistic tools (e.g., specific vocabulary, past tenses, and the content of the story)—both in language comprehension and production—in order to complete the two requirements of the writing task: (a) stating questions they had about the story due to the author's rhetorical style of not providing much detail in his writings and (b) expressing their analytic opinion about how the omission of some details in the short story helped represent the turbulent history of Colombia. The task requirements were framed within students' Zone of Proximal Development (ZPD) (Vygotsky, 1978). That is, in order to successfully complete the task, students needed to draw on the concepts and knowledge they had already internalized and could independently use while grappling with new content and concepts during interpersonal interaction. Students were able to explore those concepts with the help of more expert others or with a mediating tool such as an asynchronous blog which provides time for reflection and editing as well as a potentially more relaxed environment in which to use the L2.

5. Method

As stated previously, this study aimed to systematically investigate (a) students' perspectives, attitudes, beliefs, and motivations related to the use of a course blog for developing their L2 literacy and cultural awareness and (b) how students use the blog tasks (reading and writing) to expand their knowledge of the Spanish language while drawing on their digital literacies.

5.1 Q Methodology

In order to achieve this purpose, we used Q methodology (QM) as the main mode of inquiry (Brown, 1980; Stephenson, 1953). QM provides a framework for the study of human subjectivity (Brown, 1993) because it allows researchers to investigate viewpoints (e.g., perspectives, attitudes, beliefs, and motivations) in a systematic way. QM is based on a two-fold premise: that subjectivity is communicable and that viewpoints are always anchored in self-reference; that is, participants rank statements about a particular topic of study from their own point of view, and this is "what brings subjectivity into the picture" (Brown, 1993, p. 93). Therefore, QM can be used to analyze a range of perspectives centered around a single issue or topic.

In QM, participants are asked to sort a set of statements on a given topic. Their responses are then analyzed using statistical techniques including correlation and factor analysis. The result of a QM analysis is a set of factors that can be interpreted as perspectives and a matrix that shows to what extent each participant shows similarity to each of the identified perspectives.

QM is not a quantitative measure to explore standard meaning or individual differences. Rather, it deals with intraindividual differences in significance because differences in scores between statements are assumed to reflect differences in the amount of importance attributed to them by an individual. Therefore, operant responses—rather than operational definitions—are at issue. For this reason, the concept of external validity is not an issue in QM since there are no outside criteria for an individual's own point of view (Brown, 1980). Because of this property, it does not limit the number of people who can participate.

5.2 Participants

Fifty-four undergraduate students enrolled in four different sections of Elementary Spanish II at two private universities in the northeastern United States were invited to participate in the study. These 54 students (29 female; 25 male) were asked to complete an online Q survey about their perspectives on using a course blog for developing their Spanish language literacy and cultural awareness. Thirty-four students volunteered to participate and returned valid Q sorts based on the online Q survey. In the sections that follow, we describe the data collection process and then provide analysis and discussion.

5.3 The Q Sample of Statements in the Concourse

The Q sample for this study consisted of 28 statements selected from a larger population of items known as the "concourse" (Stephenson, 1953). Following the conventions of QM, statements were selected if they were self-referent, that is, topics, concepts, or objects about which participants could express their own opinion. In the case of the present study, statements that expressed multiple opinions about the use of an educational blog in the Spanish course were selected. These statements were drawn from a larger localized population of statements drawn from blog comments at the end of the current and previous two courses in which students were asked to reflect on their own experience using a similar blog. Thus, such reflective and subjective comments expressing perspectives, attitudes, beliefs, and motivations by learners in the same course and language program are considered "multiple and contextualized in the moment, rather than fixed, or determined by sociodemographics or other characteristics of the individual [learner]" (Eden, Donaldson & Walker, 2005, p. 414). In other words, these comments represent a localized and dynamic subjectivity that is forged and shared in a sociocultural context (Stephenson, 1978). A total of 112 statements were initially collected. Next, a thematic and content analysis of those students' reflections about their experiences using the course blog was carried out by the three researchers. Once saturation of the different themes and ideas expressed by students was reached (and after grouping, comparing, and contrasting the themes and ideas), a total of 28 Q statements were finally selected for inclusion in the online Q survey administered to the students participating in the study (see the set of Q statements in Appendix A).

5.4 Q Sorting

Q sorting is the process whereby the participants model their opinions or feelings about the issue under study. The purpose of the sorting is to "get a picture of the individual's own view of, or attitude toward, the object being considered" (Stephenson, 1967). The Q sort was conducted using FlashQ, a free application developed for conducting Q sorts on line (Hackert & Braehler, 2007). Following the conventions of Q methodology, participants were asked to read through the selected 28 statements and then sort them into two piles. They placed the statements with which they agreed with the most in one pile and the statements with which they disagreed in another pile. For statements with which the participants felt neutral, they were instructed to place them in a neutral pile. After the initial sort, the participants distributed the statements along a 9-point continuum from *most agree* to *most disagree*, as shown in Figure 2.

Figure 2
Q Sort Grid Used in the Study

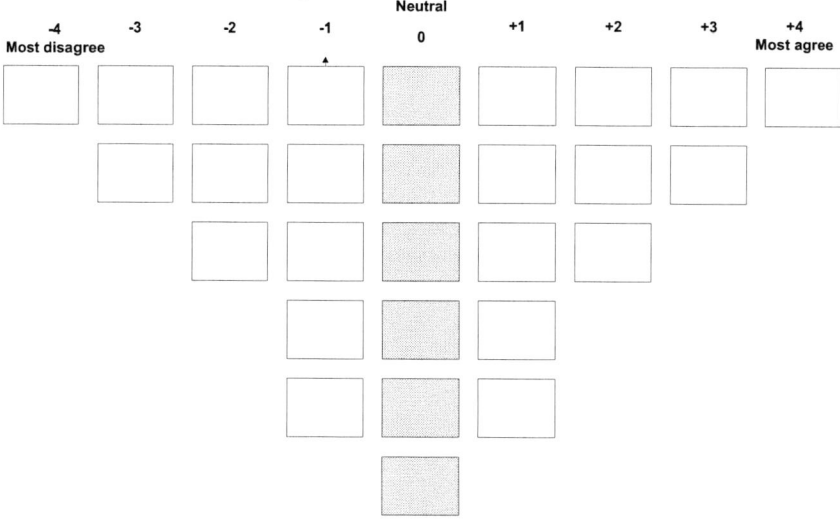

In the process of Q sorting, participants constantly compare statements with one another. The concern, therefore, is not with whether participants agree or disagree with an item, but with how they order the items in their mind (Kim & Kim, 2008). As stated earlier, Q is based on intraindividual differences in significance. Thus, when finished, participants had sorted the 28 statements into the quasinormal forced distribution.

5.5 Analyzing Data

The analysis was conducted using PQMethod (Schmolck, 2012), a dedicated computer software for the analysis of Q sort data. To analyze the data, a correlation matrix was created to compare levels of agreement among the 34 participants followed by a factor analysis on the correlation matrix to group together, as one factor, Q sorts that had similar rankings. The factor analysis and rotation used the centroid method with orthogonal rotation (Brown, 1980; Kerlinger, 1973).

For this study, three factors were identified (see the rotated factor loadings in Table 1 below). Factor loadings are assumed to be significant at the $p < .05$ level if they equal or exceed 1.96 times the standard error of a zero correlation, that is, $1.96 \times 1/\sqrt{n}$ where n is the number of items in the Q-sample (Brown, 1980). Therefore, loadings of ± 0.3704 were considered significant and indicative of a meaningful relationship between the participant and the factor type.

6. Results

Table 1 shows the three distinct factors or perspectives that were extracted from the quantitative analysis of the Q sort data (see factor loadings in Appendix B).

Table 1
Factor Correlations

Factor	% explained variance	# of sorts	1	2	3
1	11	11	1.0000	-0.1224	-0.2137
2	9	8	-0.1224	1.0000	0.2477
3	9	7	-0.2137	0.2477	1.0000

The perspective each factor represents is determined by the calculation of factor arrays. A factor array is a "Q sort configured to represent the viewpoint of a particular factor" (Watts & Stenner, 2012, p. 140). The arrays then form the basis for interpretation since they show the perspective each factor represents (see factor arrays in Appendix C).[1] Thus, following the approach proposed by Watts and Stenner, we created interpretation crib sheets for each one of the three factors. The process consists of using the factor arrays for identifying the items given the highest and lowest ranking as well as those items ranked higher or lower than any other factors in the study. Next, each factor or perspective was labeled with a descriptive name based on its salient characteristics from the positive and negative rankings. In the section that follows, we present the interpretation of each perspective with its accompanying interpretation crib sheet.

6.1 Perspective 1

Perspective 1 refers specifically to viewing the course blog as a tool for expanding students' linguistic repertoire of formal aspects of their written language instead

---

[1] Factor arrays were calculated using PQMethod. For a thorough explanation of the statistics behind this procedure, see Brown (1980, pp. 242-244).

of viewing the course blog as a space for social interaction. Factor 1 showed a perspective shared by 11 participants in the study. This perspective indicates that students believed the course blog helped them to develop their grammar, vocabulary, and more formal aspects of the language and to improve their writing. For example, students who share this viewpoint believe that writing on the blog helped them to increase their repertoire of verbal conjugations in Spanish by using more tenses and incorporating the new grammatical structures that they were studying in class into their blogs entries and comments (Statement 1, +4).[2] They also believe that writing in the blog helped them to expand their vocabulary (e.g., using new vocabulary and phrases) (Statement 3, +3) and to improve their writing fluency in Spanish (7, +3), as indicated by their positive ranking of Statements 1, 3, and 7 (shown in bold in Table 2). Clearly, this group of students perceived the blog as a tool for improving their written language and saw the blog assignment as meaningful work (10, -3). Also, students sharing this perspective affirmed that they did not expect the instructors to correct their blog work as with other typical written compositions, which is displayed by the negative ranking of Statement 27 (-3). This result may indicate a high level of autonomy as well as a high level of confidence regarding the participants' current knowledge of Spanish. Moreover, for students with Perspective 1, the more informal way of writing in the blog turned the typical individual grammar and vocabulary homework into a meaningful group activity, thus helping them to develop their formal written communication in Spanish, as suggested by the ranking of Statements 18 and 20 (+2) respectively.

Table 2
Interpretation Crib Sheet for Factor 1

Statement #	Statement	Ranking
1	**Writing on the blog has helped me to expand my repertoire of verbal conjugations in Spanish (Using more tenses; incorporating whatever we are doing in class into my blogs).**	**+4**
Items ranked higher in Factor 1 array than in other factor arrays at +4		
3	**Writing on the blog has helped me to expand my vocabulary (e.g., using new vocabulary and phrases).**	**+3**
6	I like writing for the blog because I feel that my ideas are valued whether those ideas are expressed grammatically or not.	+1
7	**The blog helped me improve my Spanish writing fluency.**	**+3**
11	The blog encouraged me to be more thoughtful (in content and structure) in writing in Spanish for a real audience.	+1
18	Writing for the blog turned my individual grammar and vocabulary homework into a meaningful group activity.	+2

---

[2] Here, the first number in parentheses (1) denotes the number of the statement, and the second number (+4) represents the final ranking for this statement in the factor.

20	Although the nature of the blog encourages a more informal way of communication, it has helped me to develop my formal communication in Spanish.	+2
24	The blog is a fun way to practice Spanish.	+2
Items ranked lower in Factor 1 array than in other factor arrays		
10	I wrote in the blog mainly to complete a course requirement.	-3
14	*The blog afforded me opportunities for personal expression in Spanish.*	-1
15	*The blog provided me with opportunities to engage in real communication in Spanish with my peers.*	-1
16	*The blog contributed to develop a sense of community in our class by allowing students to get to know each other at a personal level.*	-1
17	*The blog allowed students to freely express their opinions without worrying about the rigid structure of the formal academic writing.*	-1
27	**The blog would have been more significant if my entries had been corrected by the instructor as with regular written compositions.**	-3
28	My understanding of Hispanic cultures increased with the work on the blog.	-2
Item ranked at -4		
25	Completing blog assignments took me longer than completing workbook exercises.	-4

Furthermore, triangulation of data taken from students' reflections on the use of the course blog during the last blog task of the semester supports this group of students' views of the blog as helping them to experiment with new linguistic forms in developing more formal aspects of the language and improving their writing ability. Consider, as an example, Calvin's metacognitive reflection in the following excerpt:[3]

Excerpt 1 (Calvin, May 4, 2011, 8:36 am)
Escribiendo en el blog ha sido bien. el mejor de todo es que me ha permitido tratar los tenses y las frases que yo no puedo usar en el habla. (...) Esta frase es un ejemplo. En el blog trato usar cosas como "inf + la / lo " porque yo no lo entiendo completamente.

In Excerpt 1, Calvin sees the blog as a space for experimenting with the language (e.g., using more tenses, incorporating classroom work into blogs). He later stated that practicing new tenses and vocabulary in the blog had helped him to feel able and ready to use those new forms when speaking: *Cuando he practicado, me siento más poder de pensar rápidamente y tal vez puedo usar cosas nuevas cuando hablo.* 'When I have practiced, I feel more able to think quickly and perhaps I can use new things when I speak.'

---
[3] All names are pseudonyms.

6.2 Perspective 2

Perspective 2 refers specifically to viewing the blog as an enabling tool for students to engage in real communication and personal expression while paying attention to form.

Table 3
Interpretation Crib Sheet for Factor 2

Statement #	Statement	Ranking
22	**I could express myself in the blog in ways that I would not be able to do during regular oral class discussions.**	+4
Items ranked higher in Factor 2 array than in other factor arrays at +4		
4	**The use of proper grammar in the blog is important to me.**	+2
14	**The blog afforded me opportunities for personal expression in Spanish.**	+2
15	**The blog provided me with opportunities to engage in real communication in Spanish with my peers.**	+3
27	**The blog would have been more significant if my entries had been corrected by the instructor as with regular written compositions.**	+1
Items ranked lower in Factor 2 array than in other factor arrays		
6	I like writing for the blog because I feel that my ideas are valued whether those ideas are expressed grammatically or not.	-2
9	The fact that my classmates, and not only the instructor, also participated in the blog, encouraged me to complete the blog assignments.	-1
18	Writing for the blog turned my individual grammar and vocabulary homework into a meaningful group activity.	-2
19	Having different kinds of input (video, text, poetry, images, etc.) helped me understand authentic texts.	-1
20	*Although the nature of the blog encourages a more informal way of communication, it has helped me to develop my formal communication in Spanish.*	-3
23	*The blog made me practice Spanish outside of class.*	-3
Item ranked at -4		
13	The amount of reading in Spanish in this course increased as result of having blog assignments.	-4

Table 3 shows this perspective. Students who shared this perspective saw the blog as an enabling space where they could express themselves in ways that they could not have otherwise in regular oral interactions in the classroom (22, +4). They used the blog as affording opportunities to engage in real communication in Spanish with their peers (15, +3) and as a medium for personal expression (14, +2), while paying attention to their use of proper grammar (4, +2). These results suggest that, for these students, having more time to frame and produce their entries and comments allowed them to practice and use grammatical concepts that they did not have readily available when interacting orally in face-to-face situations in the classroom (see statements in bold in Table 3). That is, the blog's tasks provided the necessary scaffolding within this group of students' zone of proximal development for them to successfully complete such tasks using an array of new linguistic features.

The negative ranking of Statements 20 and 23 (shown in italics in Table 3) may seem somewhat contradictory to those ideas expressed through the statements in the positive rankings within this same second perspective. Here, students agreed that the blog was not a place to have more reading and overall language practice, but rather a place to use the language to communicate. This could be precisely the case of those students who had difficulty participating in oral interactions in regular face-to-face situations in class were able to express themselves in ways that they otherwise cannot in the classroom. The results of this second perspective are further supported by opinions of students such as Liz in the following excerpt:

Excerpt 2 (Liz, April 28, 2011, 8:52 pm)
Aunque hemos tenido la oportunidad a hablar todos los días durante la clase, porque estamos personas que está aprendiendo la lengua a un nivel de principiantes, puede ser difícil encontrar las palabras adecuadas para expresar lo que pensamos en el momento durante la clase. El blog nos da un espacio en que podemos expresar nuestros pensamientos y opiniones informalmente y compartir los detalles de nuestras vidas que no podemos articular. Esto es debido a que el blog nos permite a tomar el tiempo para escribir nuestros pensamientos en nuestro propio ritmo. ... el blog nos ha ayudado a sentirse cómodos articular ideas más complejas sin la presión de un ensayo formal.

Excerpt 2 shows that for students like Liz, having more time to frame their thoughts (e.g., in order to organize content, vocabulary, and grammar) is critical to being able to communicate ideas about different topics to classmates and at one's own pace. This may be due to the asynchronous nature of the digital medium of the blog as opposed to face-to-face interaction in the classroom setting.

Finally, the positive ranking of Statement 27 (+1) indicates that the students who share Perspective 2 would have liked to have the blogs corrected and to receive feedback from the instructors. This suggests that this group of students needed or expected more constant mediation and validation from their instructor for them to internalize the new grammatical concepts. This view contrasts with the higher level of linguistic confidence expressed by students in Perspective 1, where students clearly believed they had enough self-regulation over the linguistic forms and did not feel the need for the instructor's constant feedback as demonstrated by the negative ranking of Statement 27 (-3) in Table 2.

6.3 Perspective 3

Perspective 3 refers specifically to viewing the blog as a medium for real communication and interaction with authentic texts with no attention to formal aspects of the language (see Table 4).

Table 4
Interpretation Crib Sheet for Factor 3

Statement #	Statement	Ranking
	item ranked +4<	
17	**The blog allowed students to freely express their opinions without worrying about the rigid structure of the formal academic writing.**	+4
Items ranked higher in Factor 3 array than in other factor arrays		
5	The blog allowed me to communicate more freely because the instructor had a greater tolerance to simple errors than in formal written compositions.	+2
12	The blog helped practice reading in Spanish.	+2
19	**Having different kinds of input (video, text, poetry, images, etc.) helped me understand authentic texts.**	+3
25	Completing blog assignments took me longer than completing workbook exercises.	1
28	My understanding of Hispanic cultures increased with the work on the blog.	1
Items ranked lower in Factor 3 array than in other factor arrays		
1	*Writing on the blog has helped me to expand my repertoire of verbal conjugations in Spanish (Using more tenses; incorporating whatever we are doing in class into my blogs).*	-3
3	Writing on the blog has helped me to expand my vocabulary (e.g., using new vocabulary and phrases) .	-2
4	*The use of proper grammar in the blog is important to me.*	-3
7	The blog helped me improve my Spanish writing fluency.	-2
8	I gained more confidence in my ability to write because of my experience with the blog.	-1
Item ranked at -4		
2	*My use of grammar in spoken interactions has improved as a result of my work in the blog.*	-4

Participants who shared this third viewpoint saw the blog as a mediating tool for freely expressing their opinions as manifested through their high ranking of Statement 17 (+4). This group of students also believe that the course blog helped them with reading and understanding authentic texts and also learning about the culture, as shown by their ranking of Statement 19 (+3).

Moreover, students grouped into Perspective 3 did not seem to be worried about making mistakes when writing in the blog as opposed to producing a formal written composition. However, different from Perspectives 1 and 2, in which students demonstrated a clear motivation for improving their command of the L2 by practicing vocabulary and new grammatical forms, students sharing Perspective 3 used the blog mainly for communicating, interacting, and learning about culture. They did not see the blog as a tool for improving the learning of formal aspects of the language, but rather as a place to use language and grapple with authentic communicative needs. In other words, the way students in Perspective 3 used the blog seems to emulate the immersion context that learners encounter when traveling abroad where the goal is not necessarily always grammatical accuracy, but instead focusing on communication and being able to navigate the new environment through the use of the L2.

The results in Perspective 3 suggest that this group of students perceived writing in the classroom as a highly structured and formal exercise, and they therefore seemed to have used the course blog as a medium and an opportunity for writing their ideas or opinions without worrying about the rigid structure of the formal academic writing. Their negative ranking of Statements 2 (-4), 1 (-3), 4 (-3), and 7 (-2) (displayed in italics in Table 4) indicates that these students believe that the blog did not help them to expand their grammar and vocabulary, improve their writing fluency, or gain control over grammar in spoken interactions. This viewpoint shared by students having Perspective 3 is illustrated through the words of Mary in the following excerpt:

Excerpt 3 (Mary, May 4, 2011, 6:04 pm)
Creo que el blog, y el internet en general, es un gran recurso para las clases a uso. **Fue muy agradable para comunicarte con otros estudiantes no en nuestra clase.** *También me gusta* **la longitud de los blogs, lo suficiente para hacer que nuestras ideas conocidas. Las asignaciones en el blog que incluye vídeos y artículos eran mis favoritos,** *y se agrega un nivel adicional de aprendizaje.*

In this excerpt, Mary explained how having the opportunity to interact with the students from other sections of Elementary Spanish II was significant for her. Her favorite blog tasks, for example, were those in which she needed to make sense of a video or a newspaper article in the L2 and react to it in writing. Mary's comments show the importance she attributes to being exposed to diverse and authentic materials and communicative situations. Moreover, Mary's emphasis on the blog as a communicative tool was revealed when she said *me gusta la longitud de los blogs, lo suficiente para hacer que nuestras ideas conocidas.* 'I like the length of the blog (assignments), long enough for conveying our ideas.' However, the language command displayed in Mary's excerpt suggests that, for her, demonstrating control over the grammatical aspects already studied in class, such as subject-verb agreement and gender agreement between nouns and adjectives (e.g., *Las asignaciones en el blog que* **incluye** [instead of *incluyen*] *vídeos y artículos eran mis* **favoritos** [instead of *favoritas*] is not her main concern when writing to complete the assigned blog tasks. Indeed, Mary did not seem preoccupied with posting an error-free comment, but rather she wanted to express her opinion freely and probably in the same way this would have unfolded in a real interaction, had she been in the context of the target culture and given her current command of Spanish.

Furthermore, this interpretation is supported by the fact that students sharing Perspective 3 did not consider Statement 27 to be relevant at all (*The blog would have been more significant if my entries had been corrected by the instructor as with regular written compositions*) as it was not given any ranking by them in the crib sheet for Factor 3 in Table 4. The views of students having Perspective 3 clearly oppose the views about the same statement expressed by students from Perspective 2 (27, +1), who wanted more feedback and corrections from the instructors. The views related to Perspective 3 also oppose the beliefs of students from Perspective 1 (27, -3), for whom corrections may not be that necessary because they may feel more confident about their knowledge of the language.

7. Discussion

In this study, we set out to examine students' perspectives of the use of a course blog and their beliefs about how the blog's literacy-based tasks might impact their learning of Spanish language and culture. Using Q methodology as a tool for a systematic analysis of participants' subjectivities, the results show three different perspectives of students' attitudes, beliefs, and motivations regarding the role of the course blog and its significance in their language learning process. These perspectives are summarized in the next three sections.

7.1 Perspective 1

On the one hand, students in Perspective 1 focus on the language form and see the course blog as a tool to help them develop written academic language. They use the blog as an opportunity to explore more complex grammatical structures (e.g., new grammatical tenses) because the tool allows more time to develop their ideas and constitutes a less threatening environment for them to experiment with the different aspects of the language when writing in it. This is quite different from a context of impromptu face-to-face classroom interaction. These students seem to be invested in their mastery of their language and thus exert their own agency by focusing on what they want to get from the blog tasks and the course itself based on their perceived personal needs and motivations (e.g., more grammatical accuracy and writing practice), which may be totally different from the instructors' expectations.[4] That is to say, students in Perspective 1 show a greater interest in improving the formal aspects of the language, leaving the socializing and interactional aspect of the blog to the side. This is suggested by the rather surprising and unexpected results in the negative ranking of Statements 14, 15, 16, and 17

---

[4] Jiménez-Caicedo (2007) reported similar findings to those in Perspective 1 in a case study of a Latino woman in an ESOL/computer literacy class where through her agency she resisted most of the curricular expectations in the course by working on her own personal projects rather on the class-assigned tasks. However, what she did in class coherently reflected her multiple subjectivities in the different domains of her life.

(shown in italics in Table 2), which indicate that students in this perspective do not perceive the blog as a space for social interactions and more informal communication, or even as an environment for developing a sense of community with peers. Therefore, these results call into question the supposedly obvious benefits and potential of web 2.0 technology in language learning for "increased student interaction and collaboration ... in the target language" (Wand & Vásquez, 2012), which has also been reported in previous studies (e.g., Baten, Bouckaert, & Kan, 2009; Chen, 2009; Lee, 2006; Peterson, 2006).

Though somewhat surprising, this negative view from students in Perspective 1 about the blog not being a place for social interaction and informal communication corroborates the findings in a previous study by Zhu and Bu (2009), in which "only 27.9% of the students believe that working with their classmates in asynchronous online discussions has contributed significantly to the development of their communicative skills" (p. 156). Zhu and Bu explain their findings by comparing them to the results of a study on a regular face-to-face class by Conrade (1999), who concluded that first semester language students give low priority value to communicative skills when compared to students in fourth-semester courses in which communication can gain more importance. Therefore, the fact that participants from Zhu and Bu's study and those from the present study are students in the elementary language level (first and second semester, respectively) may explain the similar findings in both studies.

7.2 Perspective 2

On the other hand, students grouped in Perspective 2 see the blog as a tool for focused communication, while paying some attention to language. These students use the L2 to convey a message without having grammar as the main focus. Students in this second perspective perceive the blog as a space where they can express themselves in ways that they cannot in the classroom and as an opportunity to engage in real communication with their peers. For these students, the blog is mainly a medium for expressing what they are not able to say in face-to-face interactions or when writing their formal written compositions. The findings in this second perspective further support those of Zhu and Bu's (2009) study in which 50% of students commented that "[linguistically,] the process of composing ideas [in blogs] allows them to recycle the vocabulary and structures they have previously learned as they read and respond to their classmates" (p. 156). Cognitively, students often feel challenged by their limited and evolving linguistic skills since they often have to stop and think about what they want or need to say in order to convey ideas to their peers. Moreover, learners with Perspective 2 may be those who are more timid or less proficient students in foreign language courses. As Ducate and Lomicka (2005) argue that "[s]hy students may feel more comfortable in the relatively anonymous and equalizing environment of blogs and may be therefore more willing to share their comments with classmates" (p. 419; see also Payne, 2002).

7.3 Perspective 3

Finally, students in Perspective 3 use the blog almost exclusively as a tool for informal interaction. They write spontaneously in the blog and find that being exposed to multimodal texts is helpful in understanding the L2 and learning about culture. However, it is interesting to note that attention to language form and proper grammar use does not seem to be valued by this group. This supports findings by Kessler (2009), who discovered that when writing blog entries, students focused on the creation of meaning and were far less concerned about accuracy in the L2. Students grouped in Perspective 3 may consider writing in the classroom (e.g., formal compositions and written homework) as a rigid "passive and meaningless activity due to the lack of audience and purpose for their compositions" (Espitia & Cruz, 2013, p. 132; see also Quintero, 2008). In such cases, a blog provides users with an opportunity to express their ideas and opinions in a real communicative situation without worrying about the formalities of academic writing and the instructors' corrections. Indeed, these students believe the blog is a space for using the language communicatively to address an authentic audience rather than a tool for improving grammatical accuracy.

These findings for Perspective 3 are aligned with those of Sykes, Oskoz, & Thorne (2008) who argue that blogs afford L2 learners at the university level greater opportunities to exchange ideas for real-world purposes. Moreover, the findings related to Perspective 3 also support the notion that web 2.0 has given language teaching and learning an emphasis in L2 use rather than L2 learning (Johnson, 2004; Tu, Blocher, & Roberts, 2008; Warschauer & Grimes, 2007; Zhang, 2009). Dippold (2009) actually suggests that in order to fully engage with mediums such as blogs, participants need to abandon traditional roles and conventional writing models. Perhaps, as suggested by Thorne and Payne (2005), it is either time to lower the number of formal traditional compositions in the foreign language curriculum or time to abandon this type of assignment and use blog writing as an alternative.

Above all, the results of this study indicate that students in the three perspectives share a common trait: they exert their agency to get what they want rather than what the instructors expect as outcomes of the activities. Regardless of the specific learning goals an instructor may have in mind, the students are the ones who decide how they want to use the blog and what they want or do not want to get out of it. When learners participate in personal and collaborative learning activities such as the ones proposed in the course blog of this study, their "agency is constructed and oriented—but not determined" (Negueruela-Azarola, 2009, p. 250). In other words, blog-mediated language learning activities not only afford students an expanded audience (Gebhard, Shin, & Seger, 2011), but also provide a range of motivations and purposes for their Spanish literacy/language development. This situation may be similar to those in face-to-face class time when the instructors propose activities for students to engage in (e.g., answering a specific question, working in groups, role play, etc.), but the students are the ones who decide what aspects of the activities they want to focus on (pronunciation, vocabulary, social interaction, or simply reject the activity altogether). In the present study, the blog provided students with the flexibility to accomplish what they had in mind when completing the blog tasks. Participants in this study substantiate Pavlenko's (2007) argument that learners are individuals with wants and needs who may exert themselves and their interests by making deliberate choices in their language learning process, including the choice to resist learning or performing in the L2 in expected ways.

With regard to the limitations of this study, we recognize that, due to the nature of mixed-methods research and the sample size, the findings presented in this chapter may not be statistically generalizable to all foreign language learners. Nevertheless, we intend these findings to help instructors gain insights into the ways web 2.0 applications such as blogs can be perceived and used by students in the foreign language curriculum. We also aim to raise awareness of the need to identify students' attitudes, beliefs, and motivations.

8. Implications and Conclusion

This mixed-methods study has provided a systematic investigation of students' subjectivities about using a course blog within a multimodal literacy-based curriculum. The Q analysis, combined with qualitative descriptions of participants' work, revealed how students—at the same course level—had at least three very different perceptions of the course blog. They made their own choices regarding how to use and appropriate this sociocultural tool for negotiating meaning and for improving their command of Spanish. Thus, students exerted their individual agency based on their own attitudes, beliefs, and motivations when completing the language learning tasks proposed for the blog project. An interesting finding was the discovery that students chose to align with, resist, or ignore the initial expected outcomes that would result from the implementation of a blog project and web 2.0 technologies in the curriculum. As Polat, Mancilla, and Mahalingappa (2013) have recently noted, "the effectiveness of asynchronous discussion forums [blogs included] may be mitigated by the kind of motivation [attitudes and beliefs] L2 learners have" (p. 69). Nevertheless, we are convinced that students' motivations not only mitigate the effectiveness or impact of these tools, but if such motivations are identified early in the learning process (e.g., at the beginning of a semester or in elementary courses), such information could be used to improve learning.

Consequently, we are left with the following question: If students clearly have different beliefs and motivations about the use of blogs, as suggested by the results of the present study, how can we identify, acknowledge, and support an individual student's differences for the development of L2 literacy and language ability in order to capitalize on the affordances of web 2.0 technologies in the curriculum?

The results of the present study suggest ways in which foreign language instructors may further explore the potential of using blogs in the foreign/second language curriculum. First, instructors should monitor students' attitudes toward blogs and similar tools along with their actual command of their language in order to provide individual or group scaffolding to those students who may exhibit the traits or perspectives identified in this study. Second, the study illustrates that it is important for language instructors to broaden their expectations for individual students or different language levels regarding the use of blogs or similar web 2.0 technologies in the curriculum. Given the challenges and possibilities of today's increasingly technology-driven academic settings, the tasks we design for using web 2.0 tools in our teaching—as well as the expectations we may have of the outcomes from our students—could be (re)defined according to what we now know about students' beliefs and motivations about such technologies and emerging dimensions of literacy.

With regard to further research, we would like to stress Abraham and Williams's (2009) argument for the need to understand how participants, through their individual agency (including attitudes, beliefs and motivations), shape and reshape the discourse of communities such as blogs and forums. We encourage replication of the present study—or a study with a similar design—as a way to corroborate or challenge our findings. Based on the results of such studies, it would then be possible to design more specific language learning tasks that tap into learners' individual subjectivities and expectations of the use of web 2.0 technologies in the foreign or second language curriculum.

References

Abraham L. B., & Williams, L. (2009). Introduction: Understanding and analyzing electronic discourse. In L. B. Abraham & L. Williams (Eds.), *Electronic discourse in language learning and language teaching* (pp. 1-8). Amsterdam, The Netherlands: John Benjamins.

Abdous, M., Camarena, M. M., & Facer, B. R. (2009). MALL technology: Use of academic podcasting in the foreign language classroom. *ReCall, 21,* 76–95.

Ahearn, L. (2001). Language and agency. *Annual Review of Anthropology, 20,* 109–137.

Baten, L., Bouckaert, N., & Kan, Y. (2009). The use of communities in a virtual learning environment. In M. Thomas (Ed.), *Handbook of research on Web 2.0 and second language learning* (pp. 137v155). Hershey, PA: Information Science Reference.

Barton, D., Hamilton, M., & Ivanič, R. (2000). *Situated literacies: Reading and writing in context.* New York, NY: Routledge.

Barton, D. & Lee, C. (2013). *Language online: Investigating digital texts and practices.* London, England: Routledge.

Benito-Ruiz, E. (2009). Infoxication 2.0. In M. Thomas (Ed.), *Handbook of research on web 2.0 and second language learning* (pp. 60–79). Hershey, PA: Information Science Reference.

Bloome, D., Katz, L., Solsken, J., Willett, J., & Wilson-Keenan, J. A. (2000). Interpelations of family/community and classroom literacy practices. *Journal of Educational Research, 93,* 155–163.

Brinton, D. M., Snow, M. A., & Wesche, M. B. (1989). *Content-based second language instruction.* New York, NY: Newbury House.

Brown, S. R. (1980). *Political subjectivity.* New Haven, CT: Yale University Press.

Brown, S. R. (1993). A primer on Q-methodology. *Operant subjectivity, 16,* 91–138.

Brown, S. R. (1996). Q methodology and qualitative research. *Qualitative Health Research, 6,* 561–567.

Chapelle, C. A. (2001). *Computer applications in second language acquisition.* Cambridge, England: Cambridge University Press.

Chen, Y. (2009). *The effect of applying wikis in an English as a foreign language (EFL) class in Taiwan* (Unpublished doctoral dissertation). University of Central Florida, Orlando, FL. Retrieved from http://etd.fcla.edu/CF/CFE0002227/Chen_Yu-ching_200808_PhD.pdf

Cole, M. (1996). *Cultural psychology: A once and future discipline.* Cambridge, MA: Belknap Press.

Conrade, D. (1999). The student view on effective practices in the college elementary and intermediate foreign language classroom. *Foreign Language Annals, 32,* 494–512.

Deuschemann, M., Panichi, L., & Molka-Danielsen, J. (2009). Designing oral participation in Second Life: A comparative study of two language proficiency courses. *ReCALL, 21,* 206–226.

Dippold, D. (2009). Peer feedback through blogs: Student and teacher perceptions in an advanced German Class. *ReCALL, 21,* 18–36.

Ducate, L., & Lomicka, L. (2005). Exploring the blogosphere: Uses of weblogs in the foreign language classroom. *Foreign Language Annals, 38,* 410–421.

Eden, S., Donaldson, A., & Walker, G. (2005). Structuring subjectivities? Using Q methodology in human geography. *Area, 37,* 413–422.

Egbert, J., Huff, L., McNeil, L., Preuss, C., & Sellen, J. (2009). Pedagogy, process, and classroom context: Integrating teacher voice and experience into research on technology-enhanced language learning. *The Modern Language Journal, 93,* 754–768.

Espitia, M., & Cruz, C. (2013). Peer-feedback and online interaction: A case study. Íkala, 18, 131–151.

Gebhard, M., Shin, D. S., & Seger, W. (2011). Blogging and emergent L2 literacy development in an urban elementary school: A functional perspective. *CALICO Journal, 28,* 278–307.

Gee, J. P. (1996). *Social linguistics and literacies: Ideology in discourses* (2nd ed.). London, England: Taylor & Francis.

Gee, J. P. (2000). The new literacy studies: From "socially situated" to the work of the social. In D. Barton, M. Hamilton, & R. Ivanič (Eds.), *Situated literacies: Reading and writing in context* (pp. 180–196). London, England: Routledge.

Hackert, C., & Braehler, G. (2007). *FlashQ*. Retrieved from http://www.hackert.biz/flashq/downloads/

Harklau, L. (2002). The role of writing in classroom second language acquisition. *Journal of Second Language Writing, 11*, 329–350.

Herring, S. C. (2013). Discourse in web 2.0: Familiar, reconfigured, and emergent. In D. Tannen & A. M. Trester (Eds.), *Discourse 2.0: Language and new media* (pp. 1–25). Washington, DC: Georgetown University Press.

Hewings, A., & Coffin, C. (2007). Writing in multi-party computer conferences and single authored assignments: Exploring the role of writer as thinker. *Journal of English for Academic Purposes, 6*, 126–142.

Ito, M., Baumer, S., Bittanti, M., Boyd, D., Cody, R., & Herr, B. (2010). *Hanging out, messing around, geeking out: Living and learning with new media*. Cambridge, MA: MIT Press.

Jiménez-Caicedo, J. P. (2007). (Il)literate identities in adult basic education: A case study of a Latino woman in an ESOL and computer literacy class. *Colombian Applied Linguistics Journal, 9*, 25–43.

Johnson, A. (2004) Creating a writing course utilizing class and student blogs. *Internet TESL Journal, 10*(8). Retrieved from http://iteslj.org/Techniques/Johnson-Blogs/

Kerlinger, F. (1973). *Foundations of behavioral research*. New York, NY: Holt, Rinehart & Winston.

Kern, R. (2000). *Literacy and language teaching*. Oxford, England: Oxford University Press.

Kessler, G. (2009). Student-initiated attention to form in wiki-based collaborative writing. *Language Learning & Technology, 13*, 79–95.

Kim, J., & Kim, E. J. (2008). Theorizing dialogic deliberation: Everyday political talk as communicative action and dialogue. *Communication Theory, 18*, 51–70.

Lantolf, J. P. (1994). Sociocultural theory and second language learning: Introduction to the special issue. *Modern Language Journal, 78*, 418–420.

Lantolf, J. P., & Pavlenko, A. (1995). Sociocultural theory and second language acquisition. *Annual Review of Applied Linguistics, 15*, 108–124.

Lantolf, J. (2000). *Sociocultural theory and second language learning*. Oxford, England: Oxford University Press.

Lee, J. S. (2006). Exploring the relationship between electronic literacy and heritage language maintenance. *Language Learning & Technology, 10,* 93–113.

Lee, L. (2010). Fostering reflective writing and interactive exchange through blogging in an advanced language course. *ReCALL, 22,* 212–227.

Linstone, H. A. (1989). Multiple perspectives: Concept, applications and user guidelines. *System Practices, 2,* 307–331.

Likert, R. (1932). A technique for the measurement of attitudes. *Archives of Psychology, 140,* 1–55.

Lomicka, L., & Lord, G. (2009). Introduction to social networking, collaboration, and web 2.0 tools. In L. Lomicka & G. Lord (Eds.), *The next generation: Social networking and online collaboration in foreign language learning* (pp. 1–11). San Marcos, TX: CALICO.

Miyazoe, T., & Anderson, T. (2010). Learning outcomes and students' perceptions of online writing: Simultaneous implementation of a forum, blog, and wiki in an EFL blended learning setting. *System, 38,* 185–199.

McBride, K. (2009). Social-networking sites in foreign language classes: Opportunities or re-creation. In L. Lomicka & G. Lord (Eds.), *The next generation: Social networking and online collaboration in foreign language learning* (pp. 35–58). San Marcos, TX: CALICO.

Negueruela-Azarola, E. (2009). Blogs in Spanish beyond the classroom: Sociocultural opportunities for second language development. In L. B. Abraham & L. Williams (Eds.), *Electronic discourse in language learning and language teaching* (pp. 241–259). Amsterdam, The Netherlands: John Benjamins.

New London Group. (1996). A pedagogy of multiliteracies: Designing social futures. *Harvard Educational Review, 66,* 60-92.

Payne, J. S. (2002). Developing L2 oral proficiency through synchronous CMC: Output, working memory, and interlanguage development. *CALICO Journal, 20,* 7–32.

Pavlenko, A. (2007). Autobiographic narratives as data in applied linguistics. *Applied Linguistics, 28,* 163–188.

Peterson, M. (2006). Learner interaction management in an avatar and chat-based virtual world. *Computer Assisted Language Learning, 19,* 79–103.

Polat, N., Mancilla, R., & Mahalingappa, L. (2013). Anonymity and motivation in asynchronous discussions and L2 vocabulary learning. *Language Learning & Technology, 17,* 57–74.

Quintero, L. (2008). Blogging: A way to foster EFL writing. *Colombian Applied Linguistics Journal, 10,* 7–49.

Richardson, W. (2006). *Blogs, wikis, podcasts, and other powerful web tools for classrooms.* Thousand Oaks, CA: Corwin Press.

Ryshina-Pankova, M., & Kugele, J. (2013). Blogs: A medium for intellectual engagement with course readings and participants. In D. Tannen & M. Trester (Eds.), *Discourse 2.0: Language and new media* (pp. 183–200). Washington, DC: Georgetown University Press.

Schmolck, P. (2012). *PQMethod*. Berlin, Germany. Retrieved from http://schmolck.userweb.mwn.de/qmethod/

Schön, D. A., & Rein, M. (1994). *Frame reflection: Toward the resolution of intractrable policy controversies*. New York, NY: Basic Books.

Snow, M. A. (2001). Content-based and immersion models for second and foreign language teaching. In M. Celce-Murcia (Ed.), *Teaching English as a second or foreign language* (3rd ed.) (pp. 303–318). Boston, MA: Heinle & Heinle.

Stephenson, W. (1953). *The study of behavior: Q-technique and its methodology*. Chicago, IL: University of Chicago Press.

Stephenson, W. (1967). *The play theory of mass communication*. Chicago, IL: University of Chicago Press.

Stephenson, W. (1978). Concourse theory of communication. *Communication, 3*, 41–40.

Street, B. (1995). *Social literacies*. London, England: Longman.

Stryker, S. B., & Leaver, B. L. (1997). *Content-based instruction in foreign language education: Models and methods*. Washington, DC: Georgetown University Press.

Sykes, J., Oskoz, A., & Thorne, S. L. (2008). Web 2.0, synthetic immersive environments, and the future of language education. *CALICO Journal, 25*, 528–546.

Thorne, S. L., & Payne, J. S. (2005). Evolutionary trajectories, internet-mediated expression, and language education. *CALICO Journal, 22*, 371–397.

Thorne, S. L., & Reinhardt, J. (2008). "Bridging activities," new media literacies, and advanced foreign language proficiency. *CALICO Journal, 25*, 558–572.

Tu, C., Blocher, M., & Roberts, G. (2008). Constructs for Web 2.0 learning environments: A theatrical metaphor. *Educational Media International, 45*, 253–269.

Vygotsky, L. S. (1978). *Mind in society: The development of higher psychological processes*. Cambridge, MA: Harvard University Press.

Walker R., Davies, G., & Hewer S. (2012) Introduction to the internet. In G. Davies (Ed.), *Information and communications technology for language teachers* (Module 1.5). Thames Valley University/University of West London. Retrieved from http://www.ict4lt.org/en/en_mod1-5.htm

Wallace, M. J. (1998). *Action research for language teachers*. Cambridge, England: Cambridge University Press.

Wang, S., & Vásquez, C. (2012). Web 2.0 and second language learning: What does the research tell us? *CALICO Journal, 29,* 412–430.

Ware, P. D., & Warschauer, M. (2006). Electronic feedback and second language writing. In K. Hyland & F. Hyland (Eds.), *Feedback on ESL writing: Context and issues* (pp. 105–122). Cambridge, England: Cambridge University Press.

Warschauer, M., & Grimes, D. (2007). Audience, authorship, and artifact: The emergent semiotics of Web 2.0. *Annual Review of Applied Linguistics, 27,* 1–23.

Watts, S., & Stenner, P. (2012). *Doing Q methodological research: Theory, method & interpretation*. London, England: Sage.

Wertsch, J. V. (1985). *Vygostky and the social formation of mind*. Cambridge, MA: Harvard University Press.

Wilson, K., & Rosenberg, M. W. (2004). Accessibility and the Canadian health care system: Squaring perceptions and realities. *Health Policy, 67,* 137–148.

Zhang, J. (2009). Towards a creative social web for learners and teachers. *Educational Researcher, 38,* 274–279.

Zhu, M., & Bu, J. (2009). Chinese EFL students' perspectives on the integration of technology. *English Language Teaching, 2,* 153–162.

Appendix A

The 28 Q Statements

1. Writing on the blog has helped me to expand my repertoire of verbal conjugations in Spanish (Using more tenses; incorporating whatever we are doing in class into my blogs).

2. My use of grammar in spoken interactions has improved as a result of my work in the blog.

3. Writing on the blog has helped me to expand my vocabulary (e.g., using new vocabulary and phrases).

4. The use of proper grammar in the blog is important to me.

5. The blog allowed me to communicate more freely because the instructor had a greater tolerance to simple errors than in formal written compositions.

6. I like writing for the blog because I feel that my ideas are valued whether those ideas are expressed grammatically or not.

7. The blog helped me improve my Spanish writing fluency.

8. I gained more confidence in my ability to write because of my experience with the blog.

9	The fact that my classmates, and not only the instructor, also participated in the blog, encouraged me to complete the blog assignments.
10	I wrote in the blog mainly to complete a course requirement.
11	The blog encouraged me to be more thoughtful (in content and structure) in writing in Spanish for a real audience.
12	The blog helped practice reading in Spanish.
13	The amount of reading in Spanish in this course increased as result of having blog assignments.
14	The blog afforded me opportunities for personal expression in Spanish.
15	The blog provided me with opportunities to engage in real communication in Spanish with my peers.
16	The blog contributed to develop a sense of community in our class by allowing students to get to know each other at a personal level.
17	The blog allowed students to freely express their opinions without worrying about the rigid structure of the formal academic writing.
18	Writing for the blog turned my individual grammar and vocabulary homework into a meaningful group activity.
19	Having different kinds of input (video, text, poetry, images, etc.) helped me understand authentic texts.
20	Although the nature of the blog encourages a more informal way of communication, it has helped me to develop my formal communication in Spanish.
21	I felt more relaxed expressing my ideas on the blog.
22	I could express myself in the blog in ways that I would not be able to do during regular oral class discussions.
23	The blog made me practice Spanish outside of class.
24	The blog is a fun way to practice Spanish.
25	Completing blog assignments took me longer than completing workbook exercises.
26	The blog's topics were relevant to me.
27	The blog would have been more significant if my entries had been corrected by the instructor as with regular written compositions.
28	My understanding of Hispanic cultures increased with the work on the blog.

APPENDIX B

Factor Loadings

QSORT	1	2	3
part1	0.0054	0.5377*	-0.1562

part2	0.1626	-0.1779	-0.2299
part3	0.1133	-0.0007	-0.3804*
part4	0.4538*	-0.3087	-0.2149
part5	0.0646	0.5418*	0.0902
part6	0.4032	0.0106	-0.5668*
part7	0.4032	0.0106	-0.5668*
part8	-0.1342	0.3765*	-0.2070
part9	0.4424*	-0.2416	-0.0241
part10	0.2338	0.1299	0.1172
part11	0.1006	0.1678	-0.2612
part12	0.1499	-0.4763*	0.1228
part13	0.2968	0.1731	-0.0783
part14	0.6420*	-0.2295	-0.1010
part15	0.0784	0.3874*	-0.0486
part16	0.4071*	-0.0855	0.0432
part17	0.5096*	-0.0004	-0.1989
part18	-0.5377*	0.3946	0.1038
part19	-0.0229	0.3036	-0.1123
part20	0.1093	-0.0780	0.3253
part21	-0.0901	0.3433	0.2026
part22	0.4830*	0.2363	-0.0083
part23	-0.0763	0.4283*	0.0754
part24	0.4682*	0.1790	0.2989
part25	0.1732	-0.1880	0.3906*
part26	0.1559	0.1329	0.5611*
part27	0.4247*	-0.1524	0.1183
part28	0.0201	0.0995	0.5849*
part29	0.0379	0.0965	0.7196*
part30	0.5033	0.3902	0.3529
part31	-0.5337*	0.1061	0.2317
part32	0.0230	0.6579*	0.3316
part33	0.0230	0.6579*	0.3316
part34	0.5293*	0.0439	0.1202
% explained variance	11%	9%	9%

*factor loadings significant at the $p < .05$ level

APPENDIX C

Factor Arrays

#	Statement	Factor Arrays[5]		
		1	2	3
1	Writing on the blog has helped me to expand my repertoire of verbal conjugations in Spanish (Using more tenses; incorporating whatever we are doing in class into my blogs).	4	0	-3
2	My use of grammar in spoken interactions has improved as a result of my work in the blog.	0	0	-4
3	Writing on the blog has helped me to expand my vocabulary (e.g., using new vocabulary and phrases).	3	1	-2
4	The use of proper grammar in the blog is important to me.	-2	2	-3
5	The blog allowed me to communicate more freely because the instructor had a greater tolerance to simple errors than in formal written compositions.	0	1	2
6	I like writing for the blog because I feel that my ideas are valued whether those ideas are expressed grammatically or not.	1	-2	0
7	The blog helped me improve my Spanish writing fluency.	3	1	-2
8	I gained more confidence in my ability to write because of my experience with the blog.	1	1	-1
9	The fact that my classmates, and not only the instructor, also participated in the blog, encouraged me to complete the blog assignments.	1	-1	1
10	I wrote in the blog mainly to complete a course requirement.	-3	0	-1
11	The blog encouraged me to be more thoughtful (in content and structure) in writing in Spanish for a real audience.	1	0	0
12	The blog helped practice reading in Spanish.	0	0	2
13	The amount of reading in Spanish in this course increased as result of having blog assignments.	0	-4	-1
14	The blog afforded me opportunities for personal expression in Spanish.	-1	2	0
15	The blog provided me with opportunities to engage in real communication in Spanish with my peers.	-1	3	1
16	The blog contributed to develop a sense of community in our class by allowing students to get to know each other at a personal level.	-1	0	1
17	The blog allowed students to freely express their opinions without worrying about the rigid structure of the formal academic writing.	1	3	4
18	Writing for the blog turned my individual grammar and vocabulary homework into a meaningful group activity.	2	-2	0

19	Having different kinds of input (video, text, poetry, images, etc.) helped me understand authentic texts.	0	-1	3
20	Although the nature of the blog encourages a more informal way of communication, it has helped me to develop my formal communication in Spanish.	2	-3	-2
21	I felt more relaxed expressing my ideas on the blog.	-1	2	2
22	I could express myself in the blog in ways that I would not be able to do during regular oral class discussions.	0	4	3
23	The blog made me practice Spanish outside of class.	-2	-3	0
24	The blog is a fun way to practice Spanish.	2	-2	0
25	Completing blog assignments took me longer than completing workbook exercises.	-4	-1	1
26	The blog's topics were relevant to me.	-1	-1	-1
27	The blog would have been more significant if my entries had been corrected by the instructor as with regular written compositions.	-3	1	-1
28	My understanding of Hispanic cultures increased with the work on the blog.	-2	-1	1

[5] Factor arrays table shows the ranking of each statement in each group or perspective. For example, Statement 1 was ranked as +4 in Factor 1 and quite the opposite in Factor 3.

Chapter 5

Closing the digital divide — A framework for multiliteracy training

MALGORZATA KUREK
Jan Dlugosz University (Poland)

MIRJAM HAUCK
The Open University (UK)

Abstract

Recently the need to prepare learners for meaningful participation in technology-based activities and thus the need for digital competence has surfaced in the scholarly literature related to the learning and teaching of languages (Hubbard, 2004, 2013; Thorne & Reinhardt, 2008; McBride, 2009; Hauck, 2010).

The authors of this chapter argue that telecollaboration provides an ideal setting for learner preparation to this effect and that training should be designed in a way that allows participants to move along a continuum from informed reception of technology-mediated input through thoughtful participation in opinion-generating activities up to creative contribution. The input and output at the beginning and the end of the described continuum tend to be of a multimodal nature drawing on a variety of semiotic resources (Kress & van Leeuwen, 2001). Learners who can comfortably alternate in their roles as "semiotic responders" and "semiotic initiators" (Coffin & Donohue, in press) will reflect the success of telecollaborative training programs that take account of multimodality as a core element of digital communicative literacy skills, also referred to in the literature as new media literacy or multiliteracy.

Based on the theoretical framework of multimodal meaning making (Kress, 2000), a model for designing instruction grounded in multiliteracy is proposed. It will allow language educators to carefully guide learners through the aforementioned stages of multiliteracy skills development.

1. Introduction

Digital Competence (DC) has been acknowledged as one of the eight key competences for lifelong learning by the European Union (2006). As a so called transversal skill, it is seen as a key competence which enables learners to acquire other key competences such as languages, mathematics, learning to learn, and creativity and required by all citizens to ensure their active participation in society and the economy. Yet, it is only recently that the need to prepare learners for meaningful participation in technology-mediated activities has surfaced in the scholarly literature related to the learning and teaching of languages (Hauck, 2010; Hubbard, 2004, 2013; McBride, 2009; Thorne & Reinhardt, 2008, Reinhardt & Thorne, 2011). This need stems from the growing realization that young learners' digital proficiency tends to remain superficial and does not readily transfer across domains (Selvyn, 2009)—an issue to be explored later in this chapter.

In line with this, the authors argue that, far from being acquired through mere exposure to digital tools and applications, DC requires systematic preparation and training. They believe that such training should allow participants to move along a continuum from informed reception of technology-mediated input through thoughtful participation in opinion-generating activities and up to creative contribution of multimodal output. They further hold that instructional scaffolding along this continuum is called for with even greater urgency if digital literacy practices are to be undertaken by younger, presumed "digit native" learners in languages other than their first language (L1).

In the proposed approach to training, particular consideration is given to the fact that both the input and the output representing the beginning and the end of the afore-described continuum are usually of a multimodal nature, that is, draw on a variety of semiotic resources (Kress & van Leeuwen, 2001) or modes. Language learners who can comfortably alternate in their roles as semiotic responders and semiotic initiators (Coffin & Donohue, in press) will reflect the success of training that takes account of multimodality as a core element of digital literacy skills. The latter are also referred to in the literature as *new media literacy*, *multiliteracy*, or—when used as a pedagogical approach—*multiliteracies*.

In this chapter, then, the authors take a look at the concept of DC from the perspective of language instruction in technology-mediated environments. Section 2 below is dedicated to some introductory thoughts on the need for multiliteracy skills in an increasingly digitized world. Drawing on Reinhardt and Thorne (2011), the section also mentions existing pedagogical frameworks for developing digital L2 literacies, multiliteracies in particular. Section 3 outlines the challenges associated with acquiring digital fluency especially for young learners and their educational contexts. In section 4, a model for designing language instruction grounded in multiliteracy will be proposed drawing on the theoretical framework of multimodal meaning making (Kress, 2000). The model will be followed by a sample task sequence illustrating how it can be put into practice. Finally, section 5 draws the chapter to a conclusion with some summarizing remarks.

2. The Rise of Multiliteracy in the Wake of Technological Developments

Almost two decades ago the New London Group (1996) published their milestone manifesto *A pedagogy of multiliteracies*. Since then, the shift from print to screen has been unfolding with accelerating speed and with a profound impact on how we think, make meaning, communicate, create social bonds, and also on how we learn. The massive scale of these changes has affected individual cognition, sociocultural practices and interpersonal relations and has been widely discussed in the literature (Carr, 2010; Cope & Kalantzis, 2000; Knobel & Lankshear, 2007; Pegrum, 2009a; Reinhardt & Thorne, 2011; Selber, 2004). Unsurprisingly, these changes have also had repercussions on educational contexts and practices, opening up the doors of many classrooms to new possibilities of intercultural communication, online collaboration, content creation and dissemination. Finally, the paradigm shift has shaken up, if not redressed the educational power balance by labeling younger generations of seemingly versed technology users as digitally native and their less versed teachers as digitally migrating (Prensky, 2001).

Today the multiliteracy shift not only continues to unfold but, fuelled by what Pegrum (2009a, p. 34) terms a "move from a paradigm of scarcity to one of abundance" seems to be gaining speed still and broadening in terms of its scope. In our view, its educational impact can be boiled down to the following ten key developments, each of which deserves a thorough exploration, even though this is beyond the scope of the current chapter. Figure 1 summarizes these trends.

Figure 1
Recent Trends in Literacy Developments

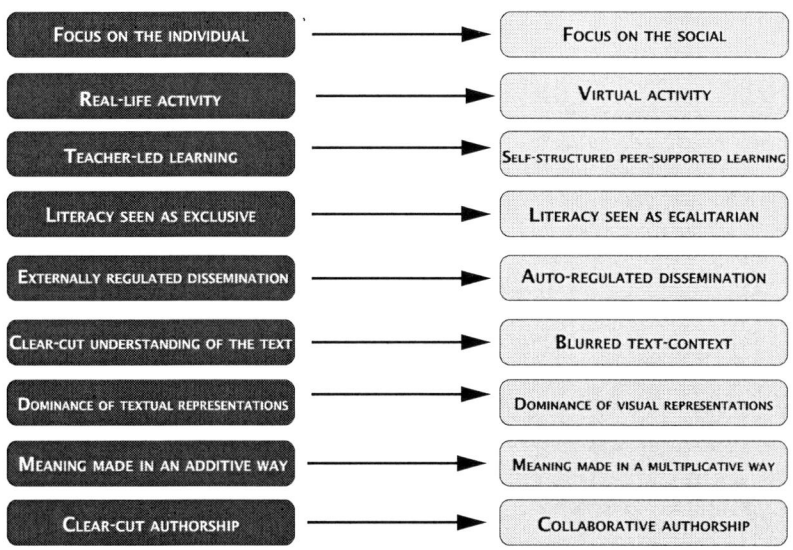

Far from being definite or complete, these developments have resulted in a transformation of how we conceptualize literacy, and they capture the directions in which daily multiliteracy practices are moving. Many of the earlier breakthrough transformations such as the interest in multimodality—defined by Kress and van Leeuwen (2001, p. 20) as "the use of several semiotic modes in the design of a semiotic product or event, together with the particular way in which these modes are combined"—have already become commonplace. This also holds true for the re-conceptualization of the meaning of 'text' and the emergence of new hybrid genres as well as the new understanding of authorship. Yet, renewed attention has recently been given to the social and cognitive dimensions of digital activity, in particular the quality of user interaction with digital content. Relevant in terms of educational demands is the realization that active participation in the changes highlighted above calls for increased personal agency and autonomy. Indeed, users who aspire to be active participants in today's multiliteracy practices need to know how to assess and evaluate their own multiliteracy experiences in contexts reaching far beyond traditional classroom walls. This is particularly relevant in technology-enhanced language learning and teaching where multiliteracy activities are—by default—mediated twice: by the technology used and by the second language (L2). One could argue though that this places L2 learning and teaching in a prime position for DC development and thus also for multiliteracy training.

The pedagogical model for learner preparation to this effect proposed further into this chapter cannot be considered separately from existing theoretical frameworks. Reinhardt and Thorne (2011) consider several pedagogical approaches for the development of digital literacies in the L2 classroom. Taking *comparisons*, one of the goal areas of the standards developed by the American Council on the Teaching of Foreign Languages (i.e., communication, cultures, comparisons, communities, and connections) as their point of reference, the authors explore functional, language awareness, and sociocultural approaches including multiliteracies, as well as their own bridging activities framework (Thorne & Reinhardt, 2008). They point out, though, that the discussed frameworks were not necessarily designed with L2 or digital literacies in mind. While some are useful to gain a more general understanding of digital media and language use that "can inform the design of pedagogical activities broadly." others are—as the authors stress—"meant to be implemented in a specific sequence" (p. 265). The concept of multiliteracies, which stems from the New London Group, falls into the latter category. At its core, it has the concepts of design and modes, including linguistic, gestural, spatial, and audio design; hence, the pluralized concept of multiliterac*ies* (Cope & Kalantzis, 2000).

An approach relevant to the conceptualization of learner training presented here comes from Hampel and Hauck (2006). Considering modes and meaning making in multimodal virtual language learning spaces, the authors propose that CMC-based language learning could also be characterized as a process of design in which "the degree of multimodal communicative competence and the degree of learner control are likely to be interdependent" (p. 11). They conclude that operating in multimodal learning environments could therefore potentially contribute to an increase in learner autonomy. Hampel and Hauck (2006) refer to Kress (2000, p. 340), who characterizes this interrelationship as follows:

> the work of design: the intentional deployment of resources in specific configurations to implement the purpose of the designers ... [T]he work of the text maker is taken as transformative of the resources and of the maker of the text. It gives agency of a real kind to the text maker.

The concepts of design and agency also underpin the increased focus on online participatory cultures and the multiliteracy skills required to join them, which in our understanding means the ability to move with ease along the continuum from informed reception of technology-mediated multimodal input through thoughtful participation in opinion-generating activities up to creative multimodal contribution.

3. Today's Educational Landscape: New Affordances Bring New Challenges

The literacy shift sketched out in the previous section provides education at the beginning of the 21st century with both opportunities and challenges, at least in those parts of the world where human beings are not affected by what Jenkins, Clinton, Purushotma, Robison, and Weigel (2006) call the "participation gap:" the fundamental inequalities in terms of access to new media technology referred to by others as the "digital divide" (see van Dijk, 2005; Warschauer, 2003). New opportunities emerge from ease of access to ever evolving tools and applications, environments, and semiotic activities, by and through which multiliterates can freely express themselves and extend their learning beyond the classroom. This abundance, however, so well reflected in the ubiquity of the prefix *multi*, requires the user to operate within increasingly multilingual, multicultural, multimodal, multigenre, and multiuser contexts. The *multis*, when combined with the rapid proliferation of available resources, channels, and modes, create an immensely rich environment that offers countless opportunities for self-expression, almost unlimited forms of meaning making, communication, and learning—both in and outside formal educational settings. Indeed, many young learners have embraced what has been termed online "participatory cultures" (Jenkins et al., 2006) and know how to build their online presence through social networking sites, avatars, audio/video casts, mash-ups, and/or by taking part in online gaming. Yet, harnessing the full potential of digital offerings requires strategic action guided by a personally unique blend of competences on a technical, cognitive, social, communicative, and even personality level. Therefore, it seems highly unlikely that multiliteracy skills for more formal educational purposes such as language acquisition can be obtained by learners through informal and uninformed technology practices. Thus, Hampel's (2006, p. 112) observation that "[w]e cannot expect learners to be competent users of the new media who are aware of the affordances and how to use them constructively" holds true more than ever before.

Indeed, while the barriers to cross-media authoring, sharing, and publishing have never been lower, the bulk of this activity, especially in case of so-called "digital native" users, falls into the remit of social affiliation and content recirculation rather than informed and purposeful content production. Beneficial as this widespread participation may be, its quality cannot be taken for granted as it is often performed as a form of social grooming reduced to brief and simplified exchanges (McBride, 2009). When it comes to on-task technology use, young learners have already been observed to display severe motivational deficiencies, following preferences for the visual—with language issues being heavily ignored—or having difficulties with staying focused or with structuring ill defined situations (for an overview, see McBride, 2009; Selvyn, 2009). Teens and university-age students, although seemingly proficient at technology, remain mostly on the operational level but are often challenged when content interpretation and independent content production come into play. What presented itself initially as digital proficiency has in many cases turned out to be a familiarity with basic affordances of the most common tools and communication modes, many of which are entertainment oriented.

In line with the above, it has been suggested that the quality of young learners' encounters with digital content tends to remain superficial, especially since digital skillfulness acquired in informal contexts does not readily transfer to formal ones (Selvyn, 2009). Sharpe (2010) points to a lack of critical and evaluative skills—a key component of digital literacy. As Littlejohn, Beetham, and McGill (2013) explain, the multitude of voices, opinions, and identities that abound on the Web in "multiple fragments, and copies and reinscriptions of themselves" (p. 134) make it difficult for learners to take a critical stance either towards online content or in opinion-generating activities unless they have experience in relevant practices. If this applies to learners' DC in their L1, it will certainly put them on an even steeper learning curve in L2 learning contexts—in both online and blended settings. Other radical differences in how today's learners process information include

- shortened attention span (Carr, 2010),
- superficial multitasking done at the expense of task engagement (Ophir, Nass, & Wagner, 2009),
- a preference for visual communication to the effect of truncated language production especially in its written form (Pegrum, 2009b),
- strong ego orientation (Thorne & Payne, 2005; Selvyn, 2009),
- familiarity with the most conspicuous affordances of technologies (Winke & Goertler, 2008), and
- use of inefficient technology solutions with focus on strategies for using specific tools rather than general exploitation ones (Hubbard, 2013).

Unsurprisingly, challenges resulting from the literacy shift do not only concern the learners but are also related to educational systems. A hierarchical and prescriptive use of technology still prevails together with a tendency to reproduce power structures known from more traditional classroom settings (Kurek & Turula, in press). This is in stark contradiction to the fact that the available technologies are, in principle, conducive to more participatory approaches.

In the following sections we present an instructional model that takes advantage of the dual mediation aspect highlighted above and that is based on the combined use of technology and L2. It can guide language educators in structuring learners' digital experiences so that those experiences turn into strategic learning choices.

4. Designing Instruction Grounded in Multiliteracy: Principles and Stages

Although the subject of dedicated learner multiliteracy training has been dealt with in the literature to a certain extent (Barrette, 2001; Hauck, 2010; Hubbard 2004, 2013; McBride, 2009; Thorne & Reinhardt, 2008), there is a growing awareness that it can improve language learning in technology-infused environments. In their review, Lai and Morrison (2013) give examples of some positive influences of learner training on learners' participation in wiki-based collaboration (Chao & Lo, 2011), autonomous listening (Romeo & Hubbard, 2013), reading comprehension skills (Zentotz, 2012), and identity construction through social networking sites (Reinhardt & Zander, 2011). Importantly, improved learner performance was achieved in many cases through awareness raising and metacognitive strategy training rather than through tool-focused instruction.

Our understanding of learner training is akin to Hubbard's (2013) — a "process aided at the construction of knowledge and skill base that enables language learners to use technology more efficiently and effectively in support of language learning objectives" (p.164). Striving for a set of universally applicable competences, we see the potential of the proposed training approach more as aiding the development of learner autonomy in digital contexts rather than tool-specific skills. Adopting such a broad perspective allows for the transfer of learning to a variety of other contexts and makes the model more adaptable to technological advancements. The overarching aim is to arrive at a pedagogical framework that initiates, scaffolds, and models learner engagement in diverse digital practices. It is hoped that training to this effect will lead to the following two outcomes: (a) a reconceptualization by the learners of the value of technology away from being primarily entertainment oriented towards its knowledge and identity-creation potential, and (b) equipped learners with a repertoire of transferable meaning-making strategies.

As the authors see it, learner interaction with multimodal content should be scaffolded around the following progressive components: reception, participation, and contribution. Following the principles of sequencing and progression, the three components of the suggested multiliteracy training are arranged according to their cognitive load, starting with informed reception, moving on to thoughtful participation and from there on to creative multimodal contribution. The choice of qualifying adjectives is intentional as they point to the nature of multiliteracy activities which, although already popular with young learners, if not modeled, tend to lack depth and direction — two highly regarded aspects of learner engagement in formal educational settings. We see these components as cumulative and complementary rather than hierarchical, with each of them preparing the ground for informed and autonomous technology use. It is the orchestration of the three stages that can, in our view, lead to fully multiliterate learner behavior.

Moreover, the learners' needs for each of the components listed above should be addressed at several levels. Although the levels are presented here in a linear order, they tend to overlap.

1. **the cognitive level**: Learners are assisted in the development of a critical stance, strategic behaviors and reflection on the processes they are engaged in.

2. **the social level**: Learners manage their online identities, develop online communities, and participate in them as well as developing collaborative and effective intercultural communication skills.
3. **the discursive level**: Learners work with different types of discourse, are asked to express and explain meaning coded in various modes as well as is being sensitized to the different aspects of offline and online communications. The discursive level, although usually taken for granted, poses a particular challenge in the L2 and its importance is additionally derived from the fact that a large proportion of online interaction is mediated by language (Pegrum, 2009b).
4. **the operational level**: Learners are prompted to move beyond the most obvious affordances of online tools and applications in the use of technology.

In our conceptualization of learner training grounded in multiliteracy we take a task-based approach following Van den Branden's (2006, p. 4) understanding of tasks as an "activity in which a person engages in order to attain an objective and which necessitates the use of language." In order to achieve comprehensive learner development, for each of the components we aim to address all the four dimensions of identified learner needs (i.e., cognitive, social, discursive, and operational. In addition, each of the components includes an element of reflective practice based on the experiential-learning premise that reflection triggers reconceptualization and change (Allwright & Hanks, 2009). Figure 2 illustrates the components of the training in question.

Figure 2
Suggested Components of Multiliteracy Training

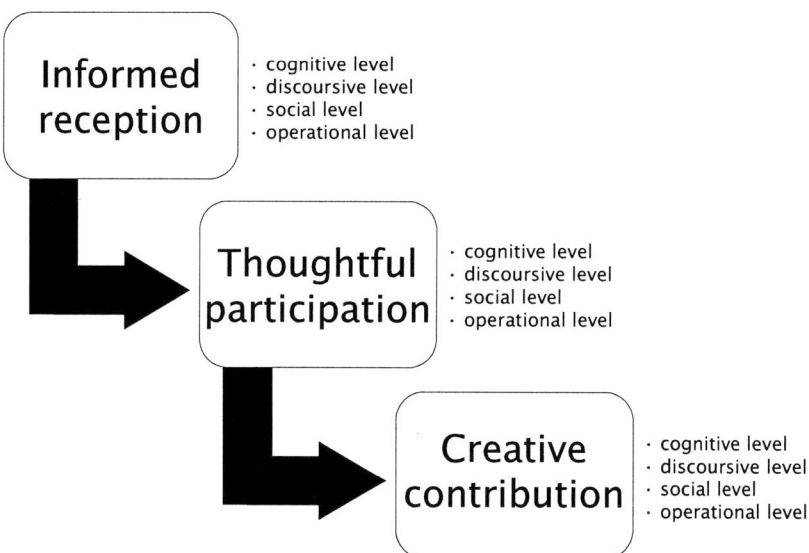

Sections 4.1-4.3 provide an overview of the three components specifying the main principles and competences in focus, and outlining suggested activities.

4.1 Component I: Informed Reception of Multimodal Input

The idea of preparing learners for informed reception of multimodal input stems from the premises of cognitive theory of multimedia learning (see Mayer & Moreno, 2003), according to which the processing demands of dealing with vast amounts of multimodal information may exceed leaner's available cognitive capacity, leading to cognitive overload and, consequently, superficial interaction with the input in question. This, as we believe, is even more complex in case of exposure to multimodal content in languages other than one's L1. We assume that careful scaffolding and modeling can lead to the reduction of cognitive load and prepare learners for situations where, as semiotic responders, they are required to attend to multimodally rich input, organize it mentally into a coherent structure and, then, apply it to new contexts.

At the stage of informed reception learners' attention should be directed towards the quality of the input, its components (e.g. the modes used), meaning potentials (Kress, 2003) and interpretative possibilities among them as well as their potential effect on the audience. This form of guidance can help learners sort through multimedia noise and notice available designs and resources in what Pennycook (2007) calls a global cacophony of voices and should lead them to the informed use of these resources in subsequent creations and contributions.

4.1.1 The cognitive level

At the cognitive level, the aim of informed reception tasks is to raise learners' awareness of context-specific affordances of available modes. We use the term affordances here in a very broad sense of the potential for communication and meaning making that particular modes hold, assuming that "visibility" of these affordances to the user comes with awareness of their existence and familiarity with their characteristic features. This approach has been suggested by Lamy and Hampel (2007), who claim that communicating and meaning making should "start by identifying the modes involved in making up a multimodal environment" and then "by considering the possibilities that they afford the learner, both as single and as combined modes" (p. 47). It is not sufficient though to ask learners to analyze different modes in isolation since, as Baldry and Thibault (2006) observe, they interconnect to make meaning and form larger patterns of discourse. Indeed, it is at this point that learners as semiotic responders should be prompted towards noticing communication codes and conventions, the possible interplay between them, and the effects to which they may work. Through the process of noticing, learners should be prompted to develop comprehension, critical analysis, evaluation, and interpretation skills. Through their increasing awareness and understanding of the available designs, they can also learn to make connections between concepts and opinions.

4.1.2 The social level

At the social level, learner preparation for informed reception embraces social multiliteracy practices reflected, for instance, in the dynamics of online communities. This can be achieved by the instructor assisting learners in filtering and synthesizing the input of many people (Warschauer & Grimes, 2007) and in first recognizing and then interpreting social relations among them. These may include among others becoming sensitive to affiliations and relations of power (Turula, 2013), cultural references, or strategies used to create a feeling of belonging and projecting one's online presence.

4.1.3 The discursive level

At the level of discourse, learners' attention should be drawn to the use of language which, if other than the L1, may serve as a nontransparent filter between user sand the meaning that they wants to communicate. This should be taken into account at the level of reception, when the learner develops preferences for meaning making modes. Language learners in particular can be tempted to choose the visual mode as a remedy for language deficiencies rather than as a result of conscious appreciation of its affordances. Therefore, tasks should be designed in such a way that they sharpen learners' senses for diverse types of linguistic discourse and discursive practices in the L2, if possible in conjunction with other online modes of expression. Diversity is understood here as access to discourses representing various cultures, genres, intentions, communication modes and language varieties. Last but not least, the reception stage should also be used to train learners in using language as an interpretation tool through which they can identify dated, biased, or exploitative resources and intentions, report and rephrase information provided visually in the form of graphs or charts, as well as communicate about other modes on a metalevel.

4.1.4 The operational level

The aim of developing informed reception on the operational level is to equip learners with technological tools for accessing, storing, organizing, and retrieving various forms of input. These should include applications for content aggregation and folksonomy services. Learners should be empowered to critically analyze the input through the lenses of the technologies used, with special attention drawn to their appropriateness and purposefulness for communicating the desired meaning.

4.2 Component II: Thoughtful Participation in Opinion-Generating Acts

There is a tendency in the current literature to treat any kind of student contribution as evidence of participatory culture and to see any type of textual production as activity reflecting multimodal creation. Yet, considering the cognitive and epistemological transformations mentioned in sections 2 and 3 above, there seems to be a strong need to help learners strengthen their linguistic production through preparation for acts of thoughtful opinion giving, especially in contexts which call for a public voice and finely tuned social competences. Although young learners are believed to be regular contributors to social networking sites, they lack experience in and preparation for discussing topics which require personal reflection and which are based on rhetorical exchanges with others. In fact, their contributions are often truncated to single words or emoticons (Pegrum, 2009b). This is confirmed by instances of learners who, when involved in telecollaborative forum or chat discussions, report lack of experience in this seemingly common forms of communication—if performed in the L2.

Contexts involving acts of opinion giving range from asynchronous emailing or text-based chat, forum contributions, blog posts, and brief evaluative comments in Web 2.0 applications to real time audiographic conferencing, with each of these activities often taking place in an intercultural and/or collaborative environment and each likely to be performed in the L2. Apart from enabling accurate self-expression on the linguistic level, thoughtful and purposeful online participation can help forge social connections and, through interaction, formulate and negotiate one's online identity (Pegrum, 2009b). Thus it is also gaining importance in online education, either small or massive scale. Operating successfully in virtual learning environments depends largely on the ability to present one's views both through individual contributions and as part of collaborative efforts. Although this seems like nothing new (similar skills are necessary in the conventional classroom) lack of physical presence and the challenges that come with mediated online presence reinforce the need for even greater clarity of thought and expression.

4.2.1 The cognitive level

On the cognitive level the aim of learner training during thoughtful participation is to raise learners' awareness of patterns and styles of involvement in diverse opinion-giving acts, as well as to sensitize them to conventions and cultural subtleties of such acts. The rhetorical decisions that learners make in the process need to be fully informed and taken with their possible interpretative outcomes in mind. Thoughtful participation also requires elements of critical thinking in the form of shifting the perspective to the other. Perspective shifting involves taking on the mental state and the cultural context of interlocutors to the effect of greater communicative efficiency. In the view of the authors, this type of sensitivity can be shaped and fostered through focused instruction.

4.2.2 The social level

The social and discursive aspect of thoughtful participation are inexorably tied together since it is mostly through language use that learners can address collaborators or communities of interest, ask for assistance, take a stance, or simply share emotions and provide support. In a practical sense this also means using the language to create a feeling of belonging and to express and nurture collaborative efforts across various contexts.

4.2.3 The discursive level

At the level of discourse, attention needs to be drawn to argumentation and negotiation skills, pragmatic competence, netiquette, and intercultural competence in the sense of being aware of culturally tinted language use. Unlike in the previous stage, learners are expected not only to interpret the meaning conveyed through input but also to articulate their own opinions by deliberately choosing and imitating a particular convention or type of discourse. This, in our view, will facilitate what Canagarajah (2005) calls "shuttling between discourses." Specifically, learners need to be trained in using a wide range of discourse functions and context-specific formulaic language. Having been sensitized at the previous informed reception stage to the ways in which linguistic features can signal emotions, positions, and stances of interlocutors, they should be prompted to use this knowledge in opinion-generating activities which, although guided by the teacher, involve communication with people from linguistically and culturally diverse backgrounds. Participation in online forums seems to be particularly well suited to the honing of thoughtful participation since, unlike traditional spontaneous face-to-face exchanges, it allows for a greater focus on linguistic form, enables careful structuring, and, thanks to the ease of archiving, fosters reflection on the language used.

4.2.4 The operational level

The operational level of participatory activities will obviously vary depending on the communication mode(s) used. While participation in forum discussions is rendered in a fairly traditional and technology transparent fashion, taking part in videoconferencing poses a greater challenge: it requires a smooth integration of input drawing on multisensory stimuli (video, textual chat, voice transmission) combined with information conveyed graphically, all of which serve as background to language production. Heavy cognitive load of videoconferencing may result in a split attention effect, that is, a situation in which attention resources need to be directed to the visual at the expense of linguistic form, a process to be observed in other cognitively complex situations such as dealing with technical problems in the L2. This can be amended for by automatizing at least some elements of the situation through pretraining or segmenting (see Mayer & Moreno, 2003).

4.3 Component III: Creative Contribution

The stage of training learners for creative contribution is supposed to assist them in both individual or joint creation of tangible artifacts and their subsequent dissemination. It is separated from thoughtful participation on the ground that it allows for a greater blend of modes and, thus, does not depend to the same extent on linguistic appropriateness. While the stage of thoughtful participation is oriented towards broadly understood linguistic competence, creative contribution does not impose restrictions on the choice of the medium, encouraging multimodal self-expression. Just as the aim of thoughtful participation is to enable the learner to shuttle between communities and discourses, the aim of creative contribution is to channel the previously acquired skills and knowledge into the act of purposeful shuttling between modes, genres, and tools as learners progress gradually from being context sensitive as semiotic responders to context transforming as semiotic initiators (Canagarajah, 2003).

4.3.1 The cognitive level

On the cognitive level, creative contribution orients learners towards the output. For instance, attention can be given to helping learners structure and express their original ideas or, more in line with the bricolage style, prompt them to approach the available input creatively and use it as canvas for their own production. In line with Cope and Kalantzis's (2000) idea of "design," this will inevitably lead to the remixing and re-purposing of the existing modes and symbols so that new meanings and new interpretations emerge. One of the possible areas for learner creativity is the restructuring of ill structured content or encoding it in a different mode. Whichever approach is taken, it is important to avoid product orientation but rather sensitize learners to the process of creation instead, with due recognition of authorship through, for example, Creative Commons licensing procedures.

4.3.2. The social level

When it comes to the social dimension of creative contribution, this can be catered for through tasks which ask learners to collaborate online on the joint production of a tangible artifact. Co-creation in particular is greatly facilitated by those Web 2.0 tools which afford the co-editing of content and, thus, foster authentic reciprocated social interactions between task participants. Through goal-oriented collaboration, learners start negotiating their online identities (Pegrum, 2009b) and, in the process, learn to recognize and fulfill their role in a team, provide support or constructive criticism to other team members, and maintain task orientation. Focused collaborative interaction in communities of practice also promotes inquiry and reflexivity—the key principles of individually perceived learning.

4.3.3 The discursive level

On the discursive level due attention should be given to the language of negotiating, analyzing, and cooperating, best realized through raising awareness of the functional context of utterances, in particular the role of pragmatics. Additionally, the language used for describing a process or technical problems needs to be foregrounded. At this stage learner attention should also be drawn to other language varieties and nonnative models in the context of social practices, with focus on how the use of particular rhetorical devices leads to the desired outcome.

4.3.4 The operational level

Paradoxically, the operational aspect of the creative contribution component seems to be of even greater relevance here than in the two previous ones. This is because the qualities of artifacts as well as the process of creation itself will depend on the choice of technologies and one's familiarity with their affordances. A great variety of available multimedia authoring tools makes it possible to address learners' unique preferences and the individual differences that inevitably exist within the digital domain. Therefore, one of the guiding principles at this stage should be presenting learners with choices rather than imposing particular technological solutions.

4.4 Multiliteracy Training: Practical Implications

The sequence presents a sample task attempting to address most of the elements of the model presented above.

Digital Biographies

The task serves as an introduction to an intercultural online exchange in which participants from two institutions in two different countries are given the opportunity to collaborate online.

Task: students prepare a multimodal presentation in order to introduce themselves to their virtual partners.
Timing: flexible
Technology: online applications suitable for various formats of personal introductions, for example, About.me, Glogster, Prezi, Windows MovieMaker

Component 1: Informed reception
Format: teacher-guided in-class activity (face-to-face)

The teacher presents examples of online presentations created with a variety of tools and by people (individuals and groups) from various cultural backgrounds.
Note for the teacher: the students should be able to access the selected examples from their own devices.

Level	Activity
cognitive	*Group discussion:* Which modes have been used to create the presentation? How do the modes used affect your impression of the author(s)/ the person (people) who created the introduction? Consider the design features that a good presentation should have.

discursive	*Group work:* How do the authors address their audience? Which communication modes do they use? *Compiling a glossary:* In your groups make a list of 5 expressions useful in analyzing and interpreting multimodal artifacts.
social	*Group discussion:* In what way do the presentations reveal cultural background information about the author(s)? Can you detect aspects that you consider different from your own cultural background?
operational	*Home assignment:* Investigate the online applications used for creating the presentations. Which of them 1) made the greatest impact on you and how do you think the impact was achieved? 2) would you like to use for your own/your group introduction and why?

Component 2: Thoughtful Participation
Format: online collaboration

Students work on line (either synchronously or asynchronously, or both) in their intercultural groups (i.e., with peers from the partner institution). They are to agree on a group name that will best reflect their group identity.

Level	Activity
cognitive	Analyzing similarities and differences.
discursive	Practicing the language of negotiation, with attention directed towards pragmatics (i.e., turntaking, making suggestions, etc.)
social	Sharing ideas with a group, paying attention to timely replies and contributions, forming relationships in the group.
operational	Mastering the technology chosen as the main group communication channel, be it a forum, a wiki or a Skype conference.

Component 3: Creative contribution
Format: Online collaboration

Using an application (or various applications) of their choice, participants work in intercultural groups to create a joint presentation of their group. They consider which representational modes allow for best possible communication of their newly created group identity including their various cultural backgrounds and the educational contexts within which they are carrying out the exchange.

Note for the teacher: if possible, students should be able to either present their products to the whole group or publish them online for other groups to see and comment on.

Level	Activity
cognitive	Deciding on the content and structure of the artifact to be produced (the presentation). Selecting the tools according to jointly established criteria.
discursive	Maintaining on-task communication
social	Online collaboration, recognizing one's role in a team, timely communication and contribution. Expressing one's opinion in a non-intrusive fashion.
operational	Students explore one or more of the tools affording synchronous or asynchronous collaboration.

Reflection:
Which part of the task did you find most challenging and why?
What other forms of creating an online presence do you have you become aware of?

The sequence of activities described above serves as an example of purposeful scaffolding of learners' experiences with multimodal input and output. The aspect of its novelty stems from its universal character and the fact that it attempts to address learner literacy needs on various levels. Similarly to what is happening in a language classroom, the learner is guided from observation of the desired acts, through their interpretation to the final performance, with the teacher gradually withdrawing support. Obviously, particular components can be used flexibly in various combinations, including a flipped order in which creative contribution serves as a trigger for students' participation in opinion-generating acts. In our view, it is the orchestration of practices representing all three components that leads to fully informed multiliteracy.

The role of the teacher deserves particular attention in this context because it evolves along the suggested continuum. At the stage of informed reception it rests by and large on providing learners with a wide variety of technology-mediated input and—through a sequence of well structured tasks—bringing into focus the various aspects of modes and meaning making and their interpretative potential. In this sense, informed reception activities serve as the first stepping stone towards autonomous learner participation in other digitally mediated practices. This orientation changes in activities focusing on thoughtful participation in which the teacher's role is to cultivate interaction and help students make informed rhetorical decisions. In the creative contribution component, teacher's role will, in addition, embrace the nurturing of various forms of learner design and creativity.

A very important aspect that remains untouched in this chapter is that learner multiliteracy preparation is a reflective practice that can be attended to either at every level of the training in question or in a final wrap-up stage of any task sequence. As pointed out in the first section of this chapter, we understand reflection as an engine for the reconsideration of knowledge, very much in line with Kohonen's (1992) view that reflection provides a bridge between experience and theoretical conceptualization. It is through reflection that learners can voice their questions, doubts and discoveries, developing their "third spaces" which "exist in the interstices between students' cultural and educational experiences and where there is ample of space for the reconstruction and reconstruction of pedagogical, linguistic and cultural knowledge and understanding" (Pegrum & Bax, as cited in Pegrum, 2009b, p. 24). This process can be triggered by asking learners to reflect on their experiences, adopt other perspectives, or come up with possible alternatives, to mention the most common techniques.

5. Concluding Remarks

Starting from the premise that technology-enhanced language learning is a particularly apt context for multiliteracy skills development we have proposed a three-tiered framework for learner training to this effect. It aims to alleviate concerns about the quality of learners' interaction with multimodal content, especially in the case of young learners. Guided by the principles of a cognitive theory of multimedia learning, we propose to scaffold learners' experiences with multimodal input and output by breaking them down into stages of growing complexity. As we believe, the commonly held assumption that today's learners are multiliterate by definition is a fallacy which, if uncorrected, will deepen the already existing divide between naive and expert users of technology. The approach is task based and draws on a theory of multimodal meaning making and its core concepts of modes and design. It distinguishes three phases during which participants move along a continuum from informed reception of multimodal input through thoughtful participation in opinion-generating acts and up to creative contributions. In doing so participants should mature in their roles as informed semiotic responders and creative semiotic initiators. At the same time, they will gradually develop the ability to explain how media shape their perception, reflect on their technology-based activities, and develop appropriate standards for their own media practices and thus overcome what Jenkins et al. (2006) have termed "the transparency problem." Tasks designed to facilitate this process should allow educators to address the rapidly increasing second level digital divide (Hargittai, 2002) (i.e., a gap resulting from the growing differences in the quality of people's online activities rather than from mere access to technology). In our view, such a gap separates those who can comfortably move along the above-mentioned continuum and who are therefore multimodally competent and thus multiliterate from those who cannot. Such competence and literacy is reflected in the users' ability "to choose, not merely with full competence within one mode [...] but with full awareness of the affordances of many modes and of the media and their sites of appearance" (Kress, 2003, p. 49). Only multiliterate users, we hold, will be able to move as semiotic responders and initiators (Coffin & Donohue, in press) along the continuum.

References

Allwright, D., & Hanks, J. (2009). *The developing learner: An introduction to exploratory practice*. Basingstoke, UK: Palgrave Macmillan.

Baldry, A., & Thibault, P. J. (2006). *Multimodal transcription and text analysis: A multimedia toolkit and coursebook*. London/Oakville: Equinox.

Barrette, C. (2001). Student preparedness and training for CALL. *CALICO Journal, 19*(1), 5–36.

Canagarajah, A. S. (2005). Shuttling between discourses: Textual and pedagogical possibilities for periphery scholars. In G. Cortese & A. Duszak (Eds.), *Identity, community, discourse* (pp. 47–68). Bern, Switzerland: Peter Lang.

Carr, N. (2010). *The shallows: What the Internet is doing to our brains*. New York, NY: Norton.

Chao, Y. C. J., & Lo, H. C. (2011). Students' perceptions of wiki-based collaborative writing for learners of English as a language. *Interactive Learning Environments, 19*(4), 395–411. Doi 10.1080/10494820903298662

Coffin, C., & Donohue, J. (in press) *A language as social semiotic approach to teaching and learning in higher education*. (Language Learning Monograph Series). Chichester, West Sussex, UK; Malden, MA: Wiley-Blackwell.

Cope, B., & Kalantzis, M. (Eds.). (2000). *Multiliteracies: Literacy learning and the design of social futures*. New York, NY: Routledge.

Joint Research Centre. *Digital Competence: Identification and European-wide validation of its key components for all levels of learners (DIGCOMP)*. Retrieved from http://is.jrc.ec.europa.eu/pages/EAP/DIGCOMP.html

European Union. (2006, December 30). Recommendation of the European parliament and of the council of 18 December 2006 on key competences for life long learning. *Official Journal of the European Union*. Retrieved from http://eur-lex.europa.eu/LexUriServ/LexUriServ.do?uri=OJ:L:200 6:394:0010:0018:en:PDF

Hampel, R. (2006) Rethinking task design for the digital age: A framework for language teaching and learning in a synchronous online environment. *ReCALL, 18*(1), 105–121.

Hampel, R., & Hauck, M. (2006). Computer-mediated language learning: Making meaning in multimodal virtual learning spaces. *JALT-CALL Journal, 2(*2),3–18.

Hargittai, E. (2002). Second-level digital divide: Differences in people's online skills. *First Monday, 7*(4). Retrieved from http://firstmonday.org/ojs/index.php/fm/article/view/942/864

Hauck, M. (2010). Telecollaboration: At the interface between multimodal and intercultural communicative competence. In S. Guth & F. Helm (Eds.), *Telecollaboration 2.0: Language, literacies and intercultural learning in the 21st century* (pp. 219–244). Bern: Peter Lang.

Hubbard, P. (2004). Learner training for effective use of CALL. In S. Fotos & C. Browne (Eds.), *New Perspectives on CALL for Second Language Classrooms* (pp.45–67). Mahwah, NJ: Lawrence Erlbaum.

Hubbard, P. (2013). Making a case for learner training in technology enhanced language learning environments. *CALICO Journal, 30*(2), 163–178.

Jenkins, H., Clinton, K., Purushotma, R., Robison, A. J., & Weigel, M. (2006). *Confronting the challenges of participatory culture: Media education for the 21st century.* Chicago: MacArthur Foundation. Retrieved from http://digitallearning.macfound.org/atf/cf/%7B7E45C7E0-A3E0-4B89-AC9C-E807E1B0AE4E%7D/JENKINS_WHITE_PAPER.PDF

Jones, R., & Norris, S. (2005). Discourse as action discourse in action. In S. Norris & R. Jones (Eds.), *Discourse in action Introducing mediated discourse analysis* (pp. 3–14). London, England: Routledge.

Knobel, M., & Lankshear, C. (2007). (Eds.). *A new literacies sampler*. New York, NY: Peter Lang.

Kohonen, V. (1992). Experiential language learning: Second language learning as cooperative learner education. In D. Nunan (Ed.), C*ollaborative language learning and teaching* (pp. 37–56). Cambridge, UK: Cambridge University Press.

Kress, G. (2000). Multimodality: Challenges to thinking about language. *TESOL Quarterly, 34*(2), 337–340.

Kress, G. (2003). *Literacy in the new media age*. London, England: Routledge.

Kress, G., & van Leeuwen, T. (2001). *Multimodal discourse: The modes and media of contemporary communication*. London, England: Arnold.

Kurek, M., & Turula, A. (in press). Digital autonomy—wishful thinking or reality? On teacher attitudes to Web 2.0 tools. In M. Dedigovic (Ed.), *Attitudes to technology in ESL/EFL pedagogy*. TESOL Arabia.

Lai, C., & Morrison, B. (2013). Towards an agenda for learner preparation in technology-enhanced language learning environments. *CALICO Journal, 30*(2), 154–162.

Lamy, M. N., & Hampel, R. (2007). *Online communication in language learning and teaching.* Basingstoke, UK: Palgrave Macmillan.

Littlejohn, A., Beetham, H., & McGill, L. (2013). Digital literacies as situated knowledge practices: academics' influence on learners' behaviours. In R. Goodfellow & M. Lea (Eds.), *Literacy in the digital university? Critical perspectives on learning, scholarship, and technology* (pp. 126–136). London, England: Routledge.

Mayer, R. E., & Moreno, R. (2003). Nine ways to reduce cognitive load in multimedia learning. *Educational Psychologist, 38*(1), 43–52.

McBride, K. (2009). Social networking sites in foreign language classes: Opportunities for re-creation. In L. Lomicka & G. Lord (Eds.), *The next-generation: Social networking and online collaboration in foreign language learning* (pp. 35–58). San Marcos, TX: CALICO.

New London Group. (1996). A pedagogy of multiliteracies: Designing social futures. *Harvard Educational Review, 66*(1), 60–90.

Ophir, E., Nass, C., & Wagner, A. D. (2009). Cognitive control in media multitaskers. In *Proceedings of the National Academy of Sciences, 106*(37), 15583–15587. Retrieved from http://www.pnas.org/content/early/2009/08/21/0903620106.full.pdf+html

Pegrum, M. A. (2009a). *From blogs to bombs: The future of digital technologies in education.* Crawley, Australia: University of Western Australia Publishing.

Pegrum, M. (2009b). Communicative networking and linguistic mashups on Web 2.0. In M. Thomas (Ed.), *Handbook of research on Web 2.0 and second language learning* (pp. 20–41). Hershey, PA: Information Science Reference.

Pegrum, M., & Bax, S. (2007, September). *Catering to diversity through asynchronous online discussion: Linking teachers across continents.* Paper presented at Diversity: A Catalyst for Innovation, the 20th English Australia Conference, Sydney, Australia 13-15 September.

Pennycook, A. (2007). *Global Englishes and transcultural flows.* London, England: Routledge.

Prensky, M. (2001). Digital natives, digital immigrants. *On the Horizon, 9*(5), 1–6.

Reinhardt, J., & Thorne, S. L. (2011). Beyond comparisons: Frameworks for developing L2 digital literacies. In N. Arnold & L. Ducate (Eds.), *Present and future promises of CALL: From theory and research to new directions in language teaching* (pp. 257–280). San Marcos, TX: CALICO.

Reinhardt, J., & Zander, V. (2011). Social networking in an intensive English program classroom: A language socialization perspective. *CALICO Journal*, 28(2), 326–344.

Romeo, K., & Hubbard, P. (2010). Pervasive CALL learner training for improving listening proficiency. In M. Levy, F. Blin, C. Siskin, & O. Takeuchi (Eds.) *WorldCALL: International perspectives on computer-assisted language learning*. New York, NY: Routledge.

Scollon, R. (2001). *Mediated discourse: The nexus of practice*. London, England; New York, NY: Routledge.

Selber, S. A. (2004). *Multiliteracies for a digital age*. Carbondale, IL: Southern Illinois University Press.

Selvyn, N, (2009). The digital native—myth and reality. *Aslib Proceedings*, 61(4), 364-379.

Sharpe, R. (2010). *Conceptualizing differences in learners' experience of e-learning: A review of contextual models. Report of the Higher Education Academy learner Difference (HEALD) synthesis project*. Retrieved from http://www.heacademy.ac.uk/resources/detail/evidencenet/Conceptualizing_differences_in_learners_experiences_of_e-learning

Thorne, S. L. (2013). Digital literacies. In M. Hawkins (Ed.), *Framing languages and literacies: Socially situated views and perspectives* (pp. 192–218). New York, NY: Routledge.

Thorne, S. L. (2003). Artifacts and cultures-of-use in intercultural communication. *Language Learning & Technology*, 7(2), 38–67.

Thorne, S. L., & Payne, J. S. (2005). Evolutionary trajectories, Internet-mediated expression, and language education. *CALICO Journal*, 22(3), 371–397.

Thorne, S. L., & Reinhardt, J. (2008). "Bridging activities," new media literacies and advanced foreign language proficiency. *CALICO Journal*, 25(3), 558–572.

Turula, A. (2013). Between deference and demeanor: The outstanding mind in online collaboration contexts. Some insights based on the five-factor model or personality traits. *Teaching English with Technology*, 2, 3-22. Retrieved from http://www.tewtjournal.org/VOL%2013/ISSUE%202/ARTICLE1.pdf

Van den Branden, K. (2006). (Ed.) *Task based language education: From theory to practice*. Cambridge, UK: Cambridge University Press.

Van Dijk, J. (2005). *The deepening divide: Inequality in the information society*. London, England: Sage.

Warschauer, M. (2003). *Technology and social inclusion: Rethinking the digital divide*. Cambridge, MA: The MIT Press.

Warschauer, M., & Grimes, D. (2007). Audience, authorship, and artifact: the emergent semiotics of Web 2.0. *Annual Review of Applied Linguistics*, 27, 1–23.

Winke, P., & Goertler, S. (2008). Did we forget someone? Students' computer access and literacy for CALL. *CALICO Journal*, 25(3), 482–509.

Zentotz, V. (2012) Awareness development for online reading. *Language Awareness*, 21, 85–100. Doi 10.1080/09658416.2011.639893

Chapter 6

Enacting Identity through Multimodal Narratives: A Study of Multilingual Students

LILIAN MINA
Indiana University of Pennsylvania (USA)

Abstract

The purpose of this chapter is to bring to light how multimodal narratives can enact multilingual students' cultural identities as these students study in a U.S. university, living in a different culture. Through a social semiotic approach to multimodality, three multimodal narratives were analyzed. Findings indicate that through the use of multiple semiotics, the three international multilingual students represented their cultural identity as international exchange students who—despite living and studying in the US—still valued the previous experiences in their home countries and the people with whom they shared those experiences. I end the chapter with a discussion of the implications of this study for EFL/ESL teachers.

1. Introduction

The ubiquity of digital media in the lives of young adults makes it almost impossible to ignore these media in the L2 classroom. Teachers' assumptions and expectations about literacy and text form should keep pace with the new literacies and text types students develop and deal with every day outside class. These new literacies and text types mandate a change in approaches to and practices of language teaching, particularly L2 teaching. Teachers of multilingual students—those whose first language is not English—need to acknowledge new ways of engaging students with L2 language, building on the literacies these students develop and practice every day.

When international multilingual students move to a new country to live and study, they find themselves struggling with identity issues (Danzak, 2011). They have learned English as a foreign language (EFL) in the confinement of classrooms in their home countries, but they find themselves required to use it in and out of class or more as a second language (ESL) in their new environments. They are challenged by novel academic requirements that may be radically different from what they have been used to in their home contexts. Additionally, they have to struggle with lots of emotional and financial problems. All these circumstances force them to negotiate their identity and redefine who they are. Since identity is socially constructed (Block, 2006; Ivanič, 1998) and dialogic (Hull & Katz, 2006), people have multiple identities that show in different contexts and discourses. Hence, Danzak argued that international students need outlets through which they can express their identity and culture. The more innovative these outlets are, the more expressive students can be, and the easier it can become to deal with their identity conflicts. Since identity conflicts may negatively impede language learning among study-abroad students (Kinginger, 2013), helping international multilingual students come to terms with their cultural identity could have positive consequences on students' language competence. Digital literacy can provide students with innovative ways to express their cultural identities through composing multimodal texts.

When students compose a multimodal essay as a form of digital literacy, they are engaged in a deep level of exploring and interpreting their own identity (Ellis, 2013). According to Jewitt (2011c), identity formation takes place through all semiotics, not only language, or as Kress and van Leeuwen (2006) contended that the individual's semiotic choices may be used to analyze their ideologies—and I extend their contention to identities. Martin (2008) elaborated on this argument and said that digital media facilitates identity development because an individual can represent that identity through creating "statements" or "multi-media objects" (p. 155). Hence, digital literacy, as Martin suggested, becomes the individual's means to "retain a hold on the shape of his/her life in an era of increasing uncertainty" (p. 156).

Despite the large number of studies that examined identity in written work, there has not been much work, if any, that has examined the enactment of a certain identity in multimodal texts (Hull & Katz, 2006). This lack of research is more striking when it comes to multilingual students. The purpose of this chapter is to bring to light how multimodal narratives can enact multilingual students' cultural identities as these students study in a U.S. university, living in a different culture. I wish to explore the question of whether composing multimodal narratives can provide international multilingual students with an opportunity to enact and display their cultural identity and whether this would boost their L2 learning. In this chapter, I discuss the relation between multimodality as a form of digital literacy and cultural identity, the social semiotic approach to multimodality (Jewitt, 2011a) used in analyzing students' multimodal texts and identities that multilingual students seemed to display in their texts. I end the chapter with a discussion of the implications of this study for EFL/ESL teachers.

2. Objective and Research Questions

The objective of this study is to bring to light how multimodal narratives can enact multilingual students' identities as these students live and study in another culture. In order to achieve this objective, the study addresses the following research questions:

1. How do international multilingual students in a U.S. university use different semiotics to represent themselves in multimodal narratives?
2. What identities do international multilingual students in a U.S. university enact in their multimodal narratives?

3. Digital Literacy and Identity

In a short historical overview, Bawden (2008) explained that the term 'digital literacy' originated from two terms: computer literacy and information literacy. The former is perceived as the ability to use computer software efficiently while the latter is more complex and includes skills such as evaluation and appreciation of information. This functional concept was challenged by Buckingham (2008a) and extended by Bawden himself. On one hand, Buckingham did not think it was fair to limit digital literacy to some basic skills of using software tools or dealing with information. On the other, Bawden proposed a more encompassing conceptualization of digital literacy beyond the combination of computer and information literacies; he extended the concept to be a hybrid of a group of literacies: visual, remix, information, hypermedia, and social-emotional literacies. His rationale for such proposal is his belief that this hybrid of literacies can be a "survival skill in the digital era" (p. 27). Such a hybrid concept makes digital literacy more of a framework within which other literacies are enacted and integrated (Lankshear & Knobel, 2008). Yet, this concept does not mean that all other literacies are enacted equally or to the same degree within digital literacy. In other words, digital literacy can be identified as an umbrella literacy under which other literacies may find a way to be reconciled.

Buckingham (2008) offered a way to reconcile other literacies within the framework of digital literacy. He strongly argued that digital literacy should exceed reading and analyzing new media to writing and producing them. Students are encouraged to produce their own multimodal texts in which they combine "written text, visual images, simple animation, audio and video material" (p. 85). The production of multimodal texts enables students to conceptualize the media they use and their life activities. Accordingly, digital literacy should be a teacher's way to link students' everyday life activities that contribute to shaping students' identities with classroom activities.

Martin (2008) extended the discussion of the role of digital media in shaping identity. He believed that digital media facilitate identity development as they put individuals in direct and constant contact with others. Individuals can represent that identity through creating multimodal texts that they share with others. Creating multimedia and multimodal texts challenges them to stay in control over their identity as they represent these identities in a different form of meaning. Such engagement with these innovative forms of text makes individuals more aware of the role of digital technologies in their development. Therefore, digital literacy becomes the individual's means to "retain a hold on the shape of his/her life in an era of increasing uncertainty" (p. 156).

Based on these conceptualizations of digital literacy and its role in developing identity, I adopt Martin's (2005) following definition of digital literacy:

> the awareness, attitude and ability of individuals to appropriately use digital tools and facilities to identify, access, manage, integrate, evaluate, analyze and synthesize digital resources, construct new knowledge, create media expressions, and communicate with others, in the context of specific life situations, in order to enable constructive social action; and to reflect upon this process. (pp. 135–136)

The continuing use of digital technologies makes digital literacy a building block in each person's identity. Digital technologies become one of the forces in play in constructing and shaping our identity. And for a person to be identified as digitally literate, they should be able to use their digital tools to understand their identity through reflections on the role of these technologies in their life. Martin's definition means that digital literacy is more than being able to use digital technologies; it means being able to reflect on how the use of technologies can help shape one's identity through creating multimodal objects that are communicated with others.

3.1 Multimodality and Identity

As different types of media become increasingly integral, even ubiquitous, in the lives of young adults, these media claim a bigger role in forming the generation's identity. The media this generation uses consist of multiple semiotics integrated together. These semiotics become the channel through which identity formation takes place, as Jewitt (2011c) has argued. Therefore, writing using semiotics like images, sound, animation, and alphabetical text is drawing from different resources for representation of the writer within a specific culture. International students move away from their own culture to go live and study in a fully different culture in which they lack the sense of commonness they used to have with groups in their culture. In order for these students to preserve their cultural identity, or their relationship to a certain culture, they represent themselves using different semiotics combined or separately. They aim at representing their cultural identity that has been long formed in one culture but now is embedded in a different one.

Rogers, Winters, LaMonde, and Perry (2010) provided a good explanation of how multimodality can be linked to identity. When students use elements of popular culture and media to produce their multimodal texts, they in reality "display their multiple subject positions" (p. 299). Students select the music, songs, and images from their culture that best show their identity. In the case of international students, there are two cultures in play. In order to determine which culture may have the bigger influence in forming these students' identities, it is important to analyze their cultural choices as represented in the semiotics they weave to create their multimodal texts. Durst (2012) rightly contended that students "weave representations of self" (p. 48) that can be a mirror of their race, ethnicity, gender, nationality, and many other features while working on multimodal projects. Students create what Danzak (2011) called "multimodal meaning" (p. 188) through a tapestry of media, modes of expression, and intertextuality of texts and voices.

All students, and international multilingual students are not an exception, have stories to tell. In order to bring international students' stories to the classroom, it is important to think of a form that would encourage them to express themselves and represent their identities. These stories will embody students' identities (Danzak, 2010), and, as students engage enthusiastically in their own stories and narratives, their identities are made apparent to the reader of these narratives. Thus, as students engage in producing multimodal narratives, they negotiate their identities. Identity negotiation is organizing and reorganizing the way an individual sees the world (Norton, 1997). Students contemplate decisions about which cultural semiotics they choose to create their multimodal narratives. Through this identity negotiation process, students' identities are enacted and revealed in their selection of semiotics.

3.2 Social Semiotic Multimodal Approach

The methods of making meaning have become different because of the development in technology and communication patterns and modes (Jewitt, 2011c). This situation mandates integrating other modes of representation than language in understanding the complex, updated ways of representation individuals now use. Similarly, Luke (2003) explained that the analysis of multimodal texts composed by students requires utilizing new methods of data analysis beyond linguistic analysis. Such new methods may include what Luke called multimethodological analysis that would enable us to decode the semiotics of these texts.

Consequently, it is imperative to embrace semiotic analysis of multimodal narratives that are composed from different semiotics. Jewitt (2011c) argued that multimodality becomes a research methodology that enables us to understand "representation and communication in a variety of fields" (p. 3).This is what Jewitt (2011a) called social semiotic multimodal. Elsewhere, Jewitt (2011b) argued that multimodal research provides researchers with fresh tools to understand the cultural resources users utilize and organize in a certain way to represent themselves. The social semiotic multimodal approach is thus suitable for this study that aims to capture how international multilingual students represented their identities through their use of different modes and semiotic systems, moving beyond linguistic representation.

According to Jewitt (2011a), the focus of the social semiotic multimodal approach is the use of various modes of expression in a particular context. Jewitt based the social semiotic multimodal approach on Kress and van Leeuwen's (2006) concept of visual realization of meaning. She strongly and enthusiastically upheld Kress and van Leeuwen's proposal that the individual's semiotic choices may be used to analyze their ideologies. They argued that interest in certain cultural resources represents the individual's relation to the events and moments they select to represent in their multimodal text.

The social semiotic multimodal approach pays most attention to users, or sign-makers, and their choices of semiotics (Jewitt, 2011a). It also emphasizes the context within which users find and choose the available semiotic sources. Hence, the available modes users choose in a given meaning-making process become more fluid and "connected to the social context of use" (p. 30). The prime focus on users and the context in this approach allows for foregrounding their agency and identity. In other words, the multimodal product becomes what Jewitt called "a window onto its user" (p. 30) or the user's representation of identity. In a recent study, Bok (2011) used a social semiotic approach in analyzing her EFL participants' multimodal fan-fiction trailers in order to unpack these students' identities and literacy practices.

In this study, I expand Kress and van Leeuwen's (2006) contention beyond ideologies to include identities. My interest is in how international multilingual students used different modes of composing in the context of a composition class in a U.S. university to express their technological literacy mostly developed in a wholly different context and culture.

4. Research Methods

4.1 Context of the Study

In the spring 2013 semester, I taught a first-year composition class for international multilingual students in a medium-sized university in the US. I had 20 students from China, Taiwan, South Korea, Venezuela, and Saudi Arabia. As part of the course requirements, students wrote a technological literacy narrative about their relation to a digital technology and how they became literate in using that particular technology. Afterwards, students had to compose a multimodal text in which they transferred their written narratives to a digital one. Students were told to create a 3-minute multimodal, digital text in which they combined different modes of expression, such as images, songs, music, text, and animation to visually represent their narratives.

Based on an earlier survey of their technological literacy and skills that I had conducted during the first week of the semester, I realized that students were familiar with different media of producing multimodal texts: PowerPoint Presentation, Prezi, MovieMaker, and Garage Band. Therefore, I decided to give students the freedom to choose the medium of production they preferred and were more familiar with and comfortable using.

4.2 Participants

Although all my students created multimodal, digital texts in that class, I chose only three students' texts to analyze for this study. I wanted to analyze these students' texts in depth to demonstrate how they used their multimodal texts to represent themselves and their cultural identity through utilizing different modes of expression and different media of production. Two of the three students were females from Taiwan, Megan and Sue, and one was a male from China, Jim (all pseudonyms). They all spoke Mandarin Chinese as their first language. The three students were exchange students who came to the US at the beginning of the 2012-2013 academic year to study for one year in the U.S. university before they returned to their countries. The three students signed the consent form required by the IRB protocol approved for this study.

5. Data Analysis

Data in this study come from three multimodal narratives produced by the three international multilingual students. Using the social semiotic multimodal approach (Jewitt, 2011a), I used Rogers et al.'s (2010) first level of the video analysis scheme. That model was based on Rose's (2007) conceptualization of youth videos. Level one of the analysis scheme aimed to descriptively analyze both the structure (organization of shots/images) and the content (visual, textual, and aural elements) among other aspects of videos. I used the analysis of these two elements as a springboard to understand students' identities by analyzing the organization and content of the multimodal narratives. I especially focused on the combination of the visual, textual, and aural text contents and what they represented in order to explore how the international multilingual students enacted a certain cultural identity (or identities) in their multimodal narratives.

Because students were given the freedom to choose the medium of production they preferred, Megan used Prezi (a free web-based presentation application) to create a presentation to which she later added background music and voice-over narration, changing it to a movie. Sue composed her text completely at Prezi whereas Jim chose to use his webcam to shoot a video of himself talking about his technological literacy. Jim used his movie editing skills to add some animations to the video later, creating a movie.

I started analyzing the data by watching each multimodal narrative several times, taking notes on the semiotics each student used, the people they talked about, and the places they showed in their projects. I also noticed the organization of scenes in each narrative. I thought about how the students represented themselves and the possible identities they seemed to enact in their multimodal narratives. I watched each narrative three times with a one-week interval between each viewing to allow for a fresh-eye watching and analysis each time.

The next step in data analysis was to apply the analytical scheme on each narrative independently in order to decode its semiotics. This step resulted in identifying subcategories of semiotics under the three categories adopted from Rogers et al.'s (2010) study. The three students used multiple semiotics to describe their technological literacy in their multimodal narratives. The visual category included still images and animation, the textual included English written text, and the aural category included background music tracks and voice-over narration.

Through the use of multiple semiotics, the international students told their narratives about their technological literacy skills with respective digital technologies. Megan started her narrative about iPod Touch from her home country, Taiwan, where she used her boyfriend's iPod Touch for entertainment. She described how her boyfriend decided to give her the iPod Touch so they could remain in touch with each other when she travelled to the US. Using classical Chinese music as the background music, Megan depicted her experience with the iPod Touch and how it changed her life, helping her to become more independent as she was away from her family. In her narrative, Megan highlighted how she used the iPod Touch to enhance her English skills while she was in the US.

Contrary to Megan's narrative that takes place in two countries, using the iPod Touch as the link between them, Sue's narrative takes place completely in her home country, Taiwan. She told the story of her passion for photography and her journey learning to use digital cameras, first alone and then with her parents as they became interested in photography as well. Through spectacular landscape pictures, Sue zoomed in on the photography experiences she shared with her parents and what they meant to her. Interestingly, Sue used an English song as the background music for her multimodal narrative even though all the narrative takes place in Taiwan.

Similar to Sue, Jim described the ways in which his early use of the social media application WeChat impacted his life as a teenager. In his movie, he faced the webcam of his laptop computer to talk about how WeChat changed him from a shy person into a social, popular one. Jim filmed his movie with his messy dorm room in the US as his background. Contrary to Megan and Sue, Jim did not use any sound or textual elements in his movie. His movie was unique in that he did not use any text or any songs or music.

Subsequently, I moved to the next step of the analysis to draw from the independent analysis of each narrative and collectively analyze and understand the identities the three students seemed to enact in their narratives. Students seemed to emphasize their temporal stay in the US, thus representing themselves as international exchange students. As exchange students, they expressed their feelings for not only people and places in their home countries, but for the experiences they had with the digital technologies they learned in their countries. The combination of feelings and technologies helped in displaying students as avid users of various digital technologies. In other words, the three students used their literacy with computer technology to describe their literacies with other technologies. They described their technological literacies using visual, textual, and aural elements from their home and host cultures. In the coming section I discuss these findings and their implications for teaching multilingual students.

6. Findings

Due to the nature of the assignment that required students to describe their experience with a single digital technology, it was not surprising to find that the three students in this study depicted their literacy in that technology. What is truly interesting is contextualizing these literacies where they originated: home countries. Students described their technological literacies pertaining to their families, friends, and places in their home countries. The three of them seemed to cherish their countries where they learned how to use the respective technologies. Even when the technology seemed to strongly impact their life in another country, as in Megan's case, the students displayed a great deal of passion for their countries, emphasizing the temporality of living in the US as international students.

The elements students used in their multimodal narratives aided in representing their cultural identity as international students who despite living and studying in the US, still valued the experiences they had in their home countries and the people with whom they shared those experiences.

6.1 Megan's Identity and Semiotic Choices

After detailing how the iPod Touch helped her to adapt to life in the US and to become more independent, Megan pointed out that her iPod Touch was very important in sharing all these experiences with her family, friends, and boyfriend back home as seen in the screencast (a video screen capture) from her movie in Figure 1.

Figure 1
Screencast from Megan's Movie

Not only does the text on this shot complement the pictures showing her contact lists on her iPod Touch, it also highlights what really mattered to Megan even as she enjoyed her independent life in the US. Connecting with people back home seemed to be pretty important to Megan. In the following shot, Megan displayed a number of personal photos taken of her in visit to multiple U.S. cities. She said that she used her iPod Touch to share her happy moments touring these cities with people in her country.

Megan's prime interest in connecting with people in her country using a technological device and digital literacy showed how she valued her cultural identity as an international student who, although enjoyed her life in another country, still perceived that life as a temporary leave away from the country where her loved ones lived.

In addition to talking about the importance of connecting with her loved ones in her home country, Megan talked extensively about using her iPod Touch to improve her English skills. Megan used a digital dictionary on her device to be able to understand and communicate better with different people on and off campus. She also said that she had used the news applications of BBC News and CNN to read in English outside her classes and to improve her vocabulary and reading skills. She even showed snapshots of these applications in her movie (see Figure 2).

Figure 2
Megan's Use of iPod Touch to Improve Her English

The way Megan talked about using her iPod Touch to improve her English skills helped in instilling her identity as an international multilingual student who came to study for a year in the US. Although her English skills were at the upper intermediate level that qualified her to take a composition course, her exposure to English all the time and everywhere may have caused her to realize that her skills needed more sharpening. That realization may have been the reason why she was keen on learning more English through different applications on her iPod Touch. Megan's experience demonstrates the value of digital literacy on two different levels. On one level, Megan realized that the device she mainly had for fun and communication could be used for literacy tasks. On the other level, she used her multimodal narrative to represent herself as an international student who learned English as a foreign language in her country but had to learn how to communicate and function in an English-as-a-second-language environment. Digital literacy and composing were combined in a recursive process in which speaking and writing better English was the motivation behind utilizing digital literacy that, in turn, was used to display Megan's confident speaking and writing in English. Megan showed her audience that her digital literacy played a significant role in not only switching from an EFL to an ESL context, but in also adjusting and fitting in the new context in the US through enhanced English skills.

Megan's choice of a traditional Chinese music track as the background of her multimodal narrative was significant in reinforcing her identity as an international student because the music created another layer of connection to her home country. As Megan continued to emphasize the importance of connecting with her family and friends in Taiwan, the use of Chinese music supported that connection. Even when she was composing in another country and in a foreign language, she remained connected to her country and culture through music. Besides music, Megan's voice-over narration, with her recognizable accent, was the second aural element that enacted her identity as an international student. The use of voice-over narration highlights Megan's confidence in her language skills; she spoke in her L2 to narrate her technological literacy despite her awareness of what could be perceived as "problems" in her speaking skills, such as accent and stress issues. Apparently, Megan was so engaged in her multimodal narrative that she may not have allowed such "problems" to stop her from speaking in her narrative.

6.2 Sue's Identity and Semiotic Choices

Unlike Megan, Sue did not talk about her life in the US at all. Although her interest in photography and her literacy using digital cameras may have allowed her to take a lot of photos in the US, she chose to describe her technological literacy experience where it started: in her home country of Taiwan. In the shot of her Prezi presentation in Figure 3, Sue talked about how she used her new digital camera to record her daily life in Taiwan. Accordingly, I expected her to show how she used her camera to record and demonstrate her daily life in the US, which was not the case, and Sue devoted all her multimodal narrative to experiences in Taiwan.

Figure 3
Sue Showing Her Use of Her Digital Camera

In fact, in the middle of the narrative, Sue shifted from her own use of the digital camera to her parents' learning photography and then to the joint adventures of photography they had as a family. Sue described the adventures in Figure 4 as "the most impressive experience with my parents," displaying her devotion to her parents. She seemed to greatly appreciate her time with them.

Figure 1.4
Sue Describing Her Experience with Her Parents

Sue stressed her relationship with her parents, a relationship that revolved around digital photography. Showing the photos they took together in Taiwan was critical in understanding her feelings towards her family and how she valued the role of digital technology in her family's life. It was as though Sue wanted to say that Taiwan is where she belonged and where she became literate in digital photography which connected her all the more to her family.

Sue did something quite interesting: she used an English song as the background music of her narrative. Although I expected to see all the written text in English since that narrative was an assignment in an English class taught, I did not quite expect to hear an English song playing in the background of a narrative that takes place entirely in Taiwan. Employing a song from the host culture in which Sue was composing her multimodal narrative about her digital literacy in her home culture in fact implanted her identity as an international student. Sue appeared to tell the audience that although she was living in a different culture, using a different language, and incorporating elements from that culture to express her feelings, her significant moments with technology occurred in her home culture with her family. Hence, even the use of western cultural elements can be used to enact an international student's cultural identity. Moreover, Sue may have intended the use of an English song to emphasize her language proficiency in that she understood and enjoyed songs in her L2.

6.3 Jim's Identity and Semiotic Choices

Similar to Megan and Sue who described their technological literacy in the context of their home country, Jim dedicated his movie to talk about the role WeChat, a Chinese texting and communication application, played in changing his personality and thus his life. Jim opened his movie by mixing the video taken in his dorm room and the skyscrapers in New York City (see Figure 5).

Figure 5
Opening Scene in Jim's Movie

Jim introduced the video welcoming his audience to the "technology era." After this impressive opening, Jim introduced himself as an international student in his university. As early as the first 20 seconds in his movie, Jim established his identity as a Chinese student in the US, making no reference to the US whatsoever in the rest of his movie.

After talking shortly about his passion for digital technology, Jim began telling his narrative about using WeChat when he was in high school in China. He proudly described how he was the first among his classmates to download the application to his cell phone to send text and voice messages to his friends. He was excited as he said that he was a pioneer in using WeChat and that students went to him asking for advice on using it. Jim enthusiastically talked about his feelings that his use of WeChat changed him from a shy person to a popular and outgoing one who could communicate easily with others. To help his audience visualize his feelings, Jim added a thumb-up clip art to his video (see Figure 6). His voice revealed his enthusiasm and pride about the change digital technology brought to his life.

Later in the movie, Jim extensively portrayed different experiences with WeChat to emphasize the message he conveyed early in the video "how technology changed my life." Jim particularly talked about the convenience WeChat brought to his life as he was able to arrange a many social gatherings and trips without having to meet with his friends and colleagues. Moreover, he gave some details about using WeChat to know more people who shared the same interests, such as basketball. Jim claimed that he found it relatively easy to communicate with people face to face because of the confidence he gained from talking to people on WeChat.

Figure 1.6
Jim Expresses His Feelings

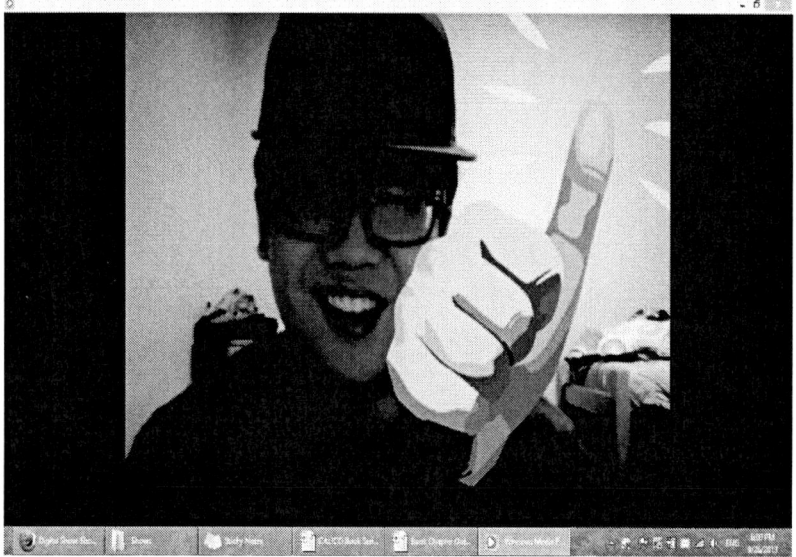

Jim's multimodal narrative helped in establishing his identity as an international student who lived in the US. The host culture was represented subtly in the background of the movie: Jim's dorm room in his U.S. university. On the other hand, his home culture was represented so explicitly and strongly in all the experiences and people Jim talked about in his narrative. Furthermore, his choice of the digital technology to describe in the narrative, a Chinese developed chat application, highlighted his identity even more clearly. Although Jim used himself as the primary visual content of his movie, his voice was the only aural content in the movie, he was able to utilize these elements effectively to represent himself and his identity. The combination of his Chinese features, noticeable Chinese accent in his speech, the Chinese technology, and the China-based experiences was very creative in representing him as an international multilingual student who from his room in the US showed his fascination with digital technology and demonstrated his technological literacy totally developed in his country. Jim appeared very comfortable and confident speaking in his L2 about his experiences that were contextualized in his L1, thus giving us more evidence that investing in digital literacy with international multilingual students can help them have a better experience learning and enhancing their L2. Jim's engagement in his multimodal narrative was easy to notice and feel.

7. Discussion

The first research question this study addressed was about the semiotics international multilingual students use to represent their identities in their multimodal narratives. Findings of the study show that the three students in this study used varied cultural semiotics to display their cultural identity as international students who temporarily live and study in the US. The visual and aural semiotics from their home cultures clearly outnumbered and dominated those from the host culture. This finding supports Rogers et al.'s (2010) contention that students choose the cultural elements that best represent themselves. Resorting more to resources from their home cultures reveals that the home cultures have a stronger influence on these students than the host culture.

The use of an English song in Sue's case could be due to the power of western mass media and culture as Kress and van Leeuwen (2006) proposed. The easy access to western media is a phenomenon that resulted from globalization. According to Buckingham (2008b) identity formation is affected by globalization that may create "a sense of fragmentation and uncertainty" (p. 1). In the case of international students who speak English as a foreign language, this sense of uncertainty can be deeper as they become detached from their culture and have to use their foreign language to survive the day-to-day activities. These students may not think about their identity until they find themselves in situations in which their identity needs to be asserted. In Sue's case, for instance, she used a western song to accompany her narrative that was comprehensively contextualized in her home culture. This finding is in accord with Buckingham's argument that new media provide young people with resources they can use to express their unique identities that may be in contrast with the surrounding circumstances or the attempts to enforce a different identity. Sue may have used the western song from her surrounding context to contrast her home context and thus affirmed her cultural identity.

Throughout the three multimodal narratives, the three students explicitly and implicitly enacted and displayed a certain cultural identity: international students who speak English as a foreign language and live and study in the US for the period of their exchange program. This finding answers the second research question posited in this study. Students were able to enact their cultural identities by using new technologies to create their multimodal narratives. Weber and Mitchell (2008) strongly argued that new technologies can offer us novel opportunities to reexamine a variety of concepts, one of which is identity. The way younger generations use new media technologies can "serve as a model for identity processes" (p. 27). As students use and interact with new digital technologies, identities can be formed and tested before they can ultimately be displayed.

This understanding of the relation between digital technology and identity formation brings us back to Buckingham's (2008a) concept discussed earlier in this chapter that digital literacy should include producing media as a conceptualization of life activities and situations. In this study, students were able to conceptualize their life experiences with different digital technologies in the context of their home countries. Subsequently, they could reflect on these experiences and realize what such experiences meant to them. This reflection evoked students' feelings of pride, love, and missing their loved ones. This show of emotions in the multimodal narratives confirms Bawden's (2008) claim that digital literacy can be the framework of reconciling disparate literacies. In this study students integrated visual literacy as they chose the images and animations in their multimodal narratives and remixed old and new cultural elements to express their emotional feelings about their cultures.

8. Implications for Teaching

The findings in this study shine light on the significance of digital literacy not only in displaying identities, but in enacting them as well. Teachers of multilingual students are thus encouraged to incorporate digital literacy in their teaching as a way to help students understand, test, and come to terms with their unique cultural identities. Although this study took place in a U.S. university, composing multimodal texts can help multilinguals across cultures and contexts attain control over their identity (Martin, 2008).

This study also pointed out the role of personal writing in negotiating and enacting certain cultural identities. Linking students' life activities with school ones is recommended if teachers are keen on making their students think and reflect on their everyday experiences. With the paramount and ubiquitous use of digital technologies, they form an integral part of students' lives and experiences. Not only does recalling these experiences in class hone in on students' digital literacy, it also employs digital literacy to enhance other literacies. Multilingual students in this study were not afraid of speaking and writing in their foreign language as they composed their multimodal narratives because their attention was primarily focused on utilizing and describing their digital literacy.

Another important implication for teachers of immigrants and international students is to help students embrace their home cultures. Many teachers unintentionally push students very hard to discard their rich experiences in their home countries in favor of the new country with its culture and language. Teachers assume that this will help students improve and accelerate their English skills. This approach is inclined to create an identity conflict for students. In order to avoid such conflict that may impede students' learning, teachers should ask students to recall their experiences from their home countries and use cultural elements from their cultures. Despite the small sample size of this study, it was obvious that students used cultural semiotics from their home countries in their English multimodal narratives. Stripping students of their cultural capital would have more negative than positive impacts on their English education in an EFL or ESL context.

9. Conclusion

This study aimed to explore the identities international multilingual students may enact and display in their multimodal narratives and the semiotic choices these students make as they compose digital texts. Findings show that digital literacy should no longer be blocked from classes because it is becoming increasingly integral in younger generations' identity, particularly cultural identity. Asking students to draw from their digital literacy outside class and from the variety of semiotics resources they have access to can help them enact their identities as they move from the comfort zone of their countries to study and live in other countries where they have to use their foreign language more frequently and for more varied purposes than literacy purposes. Allowing digital natives to exploit their digital literacy in producing multimodal texts about their personal experiences can open the door to embrace and cherish multilinguals' cultures as an indispensible component of their identity.

References

Bawden, D. (2008). Origins and concepts of digital literacy. In C. Lankshear & M. Knobel (Eds.), *Digital Literacies* (pp. 17–32). New York, NY: Peter Lang Publishing.

Block, D. (2006). Identity in applied linguistics. In T. Omoniyi & G. White (Eds.), *Sociolinguistics of identity* (pp. 35–49). London: Continuum.

Bok, E. (2011). *Exploring millennial popular culture: Multilingual adolescents' literacy and identity work in online spaces* (Unpublished doctoral dissertation). University at Buffalo, State University of New York, Buffalo, NY.

Buckingham, D. (2008a). Defining digital literacy — What do young people need to know about digital media? In C. Lankshear & M. Knobel (Eds.), *Digital literacies* (pp. 73–89). New York, NY: Peter Lang Publishing.

Buckingham, D. (2008b). Introducing identity. In D. Buckingham (Ed.), *Youth, identity, and digital media* (pp. 1–22). Cambridge, MA: The MIT Press.

Danzak, R. L. (2011). Defining identities through multiliteracies: EL teens narrate their immigration experiences as graphic stories. *Journal of Adolescent & Adult Literacy, 55*(3), 187–196. doi: 10.1002/JAAL.00024

Durst, P. (2012). *Multimodal composition and electracy: Pedagogical relays*. (Unpublished doctoral dissertation). Washington State University. Pullman, WA.

Ellis, E. (2013). Back to the future? The pedagogical promise of the (multimedia) essay. In T. Bowen & C. Whithaus (Eds.), *Multimodal literacies and emerging genres* (pp. 37–72). Pittsburgh, PA: University of Pittsburgh Press.

Hull, G. A., & Katz, M.-L. (2006). Crafting an agentive self: Case studies on digital storytelling. *Research in the Teaching of English, 41*(1), 43–81.

Ivanič, R. (1998). *Writing and identity: The discoursal construction of identity in academic writing*. Amsterdam, the Netherlands: John Benjamins Publishing Company.

Jewitt, C. (2011a). Different approaches to multimodality. In C. Jewitt (Ed.), *The Routledge handbook of multimodal analysis* (pp. 28–39). New York, NY: Routledge.

Jewitt, C. (2011b). An introduction to multimodality. In C. Jewitt (Ed.), *The Routledge handbook of multimodal analysis* (pp. 14–27). New York, NY: Routledge.

Jewitt, C. (2011c). Introduction: Handbook rationale, scope and structure. In C. Jewitt (Ed.), *The Routledge handbook of multimodal analysis* (pp. 1–7). New York, NY: Routledge.

Kinginger, C. (2013). Identity and language learning in study abroad. *Foreign Language Annals, 46*(3), 339–358.

Kress, G., & van Leeuwen, T. (2006). *Reading images: The grammar of visual design*. New York, NY: Routledge.

Lankshear, C., & Knoble, M. (2008). Introduction: Digital literacies—Concepts, policies and practices. In C. Lankshear & M. Knoble (Eds.), *Digital literacies: Concepts, policies and practices* (pp. 1–16). New York, NY: Peter Lang Publishing.

Luke, C. (2003). Pedagogy, connectivity, multimodality, and interdisciplinarity. *Reading Research Quarterly, 38*(3), 397–403.

Martin, A. (2005). DigEuLit—A European framework for digital literacy: A progress report. *Journal of eLiteracy, 2*, 130–136. Retrieved from http://www.jelit.org/

Martin, A. (2008). Digital literacy and the "digital society." In C. Lankshear & M. Knobel (Eds.), *Digital literacies* (pp. 151–176). New York, NY: Peter Lang Publishing.

Norton, B. (1997). Language, identity, and the ownership of English. *TESOL Quarterly, 31*(3), 409–429.

Rogers, T., Winters, K.-L., LaMonde, A.-M., & Perry, M. (2010). From image to ideology: Analysing shifting identity positions of marginalized youth across the cultural sites of video production. *Pedagogy: An International Journal, 5*(4), 298–312. doi: 10.1080/1554480X.2010.509473

Rose, G. (2007). *Visual methodologies: An introduction to researching with visual materials*. Thousand Oaks, CA: SAGE.

Weber, S., & Mitchell, C. (2008). Imaging, keyboarding, and posting identities: Young people and new media technologies. In D. Buckingham (Ed.), *Youth, identity, and digital media* (pp. 25–47). Cambridge, MA: The MIT Press.

Chapter 7

Digital games as practices and texts: New literacies and genres in an L2 German classroom

JONATHON REINHARDT
CHANTELLE WARNER
KRISTIN LANGE
University of Arizona (USA)

Abstract

In response to discussions on the use of commercial games as L2 learning resources, we developed a project to explore game-enhanced, literacies-informed instruction in a college-level, fifth-semester L2 German classroom. Faced with an already highly effective genre-based curriculum, we were led to consider where game literacies and genre approaches might coincide theoretically and pedagogically. This resulted in the development of a 2-week unit on gaming in which games were treated as both text and practice, an approach that aligns with a New Literacies *Design* metaphor (New London Group, 1996; Kern, 2000), with the dual objective of raising genre awareness and developing game literacies (Kringiel, 2012; Squire, 2008b; Zimmerman, 2007). The pilot implementation, reported here, had mixed reactions from learners: while some embraced gaming as a new, effective, and even pleasurable way of learning, others demonstrated resistance or skepticism because of what we interpret as a clash between expectations about language learning, play, and the constraints of the classroom and the chosen games.

1. Introduction

Although digital games and the practice of digital gaming have been of interest to CALL professionals since their inception in the 1980s (Phillips, 1987), they have recently gained increased attention (e.g., Cornillie, Thorne, & Desmet, 2012; Reinders, 2012; Sykes & Reinhardt, 2013; Thorne, Black, & Sykes, 2009). As growing numbers of L2 learners play digital games outside the classroom, and games are produced in an expanding variety of game genres and languages, it has become easier to imagine digital games and gaming as authentic, consequential, and widely applicable L2 learning resources.[1] Instructors observe that well designed games are able to motivate and engage players by introducing them to captivating narratives and communicative complexity beyond what many traditional learning activities are able to achieve.

Concerning the potential benefits of digital games, researchers and instructors have found pedagogical inspiration in research by literacy and educational gaming scholars (e.g., Squire, 2008a; Steinkuehler, 2005). In particular, James Gee (e.g., 2003, 2004, 2007) has articulated the educational capacity and instructional promise of digital games, although his work is mostly theoretical and does not specifically address L2 teaching and learning. However, Gee's work on gaming aligns with his collaborative work in the New London Group (NLG). In 1996, this group of education scholars published a groundbreaking pedagogical framework based on New Literacies (e.g., Gee, 1992; Street, 1995; for a recent history, see Barton, 2007), a reconceptualization of literacy as meaning-making practices situated in broader social, institutional, and historical relations (see also Cope & Kalantzis, 2000; Kress, 2003). For nearly two decades, the NLG's *Design* framework has provided a productive and practical heuristic for implementing language instruction based on the metaphors of literacies as social practice and learning as transformation (within foreign language education, see Kern, 2000). The framework has also informed game-based pedagogies in other fields, like Squire's (2008a, 2008b) "gaming as designed/designing experience" approach, which he has applied to secondary school social sciences instruction. Inspired by this work, scholars of technology and L2 pedagogy have begun to develop literacies-informed frameworks for digital game-mediated L2 teaching and learning contexts (e.g., Reinhardt & Thorne, 2011; Sykes, Reinhardt, & Thorne, 2010; Thorne & Reinhardt, 2008).

In response to this burgeoning discussion, we developed a project to explore game-enhanced, literacies-informed instruction in a college-level, fifth-semester L2 German classroom. Faced with an already highly effective genre-based curriculum, we were led to consider where game literacies and genre approaches might coincide theoretically and pedagogically. We therefore developed a 2-week unit on gaming in which games were treated as both text and practice, an approach that aligns with a New Literacies *Design* metaphor, with the objective of raising both genre awareness and developing game literacies. The pilot implementation, reported here, had mixed reactions from learners: while some embraced gaming as a new, effective, and even pleasurable way of learning, others demonstrated resistance or skepticism because of what we interpret as a clash between expectations about language learning, play, and the constraints of the classroom.

This chapter provides further elaboration on the idea of game literacies as digital literacies, with attention given to how they might be understood vis-à-vis L2 teaching and learning. Digital gaming as a literacy practice can be understood as play with and within a system for the purpose of meaning making that can be

---

[1] We use the terms *gaming* and *digital gaming* synonymously throughout the chapter.

developed by asking "how playing, understanding, and designing games all embody crucial ways of looking at and being in the world" (Zimmerman, 2007, p. 30). This leads to discussion of how we integrated game literacies activities into the genre-based pedagogy we used in our application. We follow with description of the unit and the results of the pilot implementation from the perspective of student-players, using their completed coursework as data sources. We then discuss implications for future iterations.

2. Background

2.1 Digital Literacies

The pluralization of *literacy* seen in terms such as *multiliteracies* and *digital literacies* marks a reconceptualization of the traditional concept of literacy as the functional ability to read and write. Instead, literacy is understood as a nexus of practices involving socially, historically, and contextually situated meaning making, shifting and contested (Street, 1984). Literacy practices are multifarious and associated with complex ecologies of community, codes, register, genre, and discourses (Gee, 1996). Literacies are thus emergent and contextually contingent; not checklists of skills or abilities, but ecologically valid because they are recognized as semiotic in nature by others who practice them (Lankshear & Knobel, 2006). This poststructural quality highlights identity, agency, and power and contrasts with definitions of literacy based on knowledge hierarchies, standards, and hegemonies.

In the reconceptualization of literacy as a plurality of socially mediated meaning-making practices, often referred to as New Literacy (Gee, 1991; Street, 1995, 2003), digital contexts have often played a central role because of the rapidly shifting, decentralized, extralocal, and multifarious nature of internet-mediated communication. The National Council of Teachers of English (2013) identifies digital literacies as *twenty-first century* literacies, noting that because "technology has increased the intensity and complexity of literate environments, the 21st century demands that a literate person possess a wide range of abilities and competencies, many literacies" (p. 1). At the same time as they are situated, however, digital literacies are also portable across contexts and participatory in nature. Jenkins, Purushotma, Clinton, Weigel, and Robinson (2006) note that they develop through a variety of social and cognitive activities, including play, simulation, performance, appropriation, judgment, navigation, negotiation, and networking. Digital literacy can be understood to include both the adept and aware use of a variety of communication technologies and the attendant discourses by which users make sense out of these practices.

Applying the concept of digital literacies to L2 teaching and learning requires an acknowledgement of language learning as emerging from, and thus inseparable from, language use in socially meaningful contexts. Reinhardt and Thorne (2011) discuss a variety of pedagogical frameworks applicable to L2 digital literacies, including media literacy (e.g., Buckingham, 2003), language awareness (e.g., Bolitho et al., 2003), genre awareness (e.g., Hyland, 2001), and the NLG's *Design* approach. Based on these approaches, Reinhardt and Thorne propose a model for *bridging activities*, activities that integrate vernacular, technology-mediated practices situated in learners' L1 primary discourses in such a way that they may bridge to new practices in the L2 (both vernacular and academic), thus fostering critical language awareness, metalinguistic and analytical skills, and establishing personal agency and relevance (Thorne & Reinhardt, 2008). Of the various digital practices cited by Thorne and Reinhardt, those associated with gaming environments are arguably among the most complex and the most ubiquitous; and yet, the majority of college students navigate these multiple levels of discourse and semiosis on a regular basis.[2]

2.2 Game Literacies

Game literacies are the literacies one develops by playing digital games and engaging in game-related social practices. Game literacy is not synonymous with being good at games, in the sense of frequently winning, although in many cases the development of game literacy might make one a more competitive player. Rather, game literacy entails awareness of games as designed representational systems involving play—both in the more colloquial sense of fun and also in the developmental sense (Vygotsky, 1978; see also Cook, 2000).[3] There are several conceptualizations of game literacies, from functional and explicit understandings to more socially informed frameworks. Kringiel (2012) promotes the concept as a response to societal criticisms, which themselves often inform pedagogical notions, that digital gaming is antisocial and addictive. He positions game literacies within Baacke's (2004) media literacy framework, which is comprised of four components: criticism, understanding, use, and production (for an explanation of Baacke's framework, see Kringiel, 2012). *Critical game literacy* is developed through metaknowledge of the logic, goals, and strategies of games and the games industry, while *game understanding* involves knowledge of the genres, structures, and aesthetic conventions in games, that is, "ability to play video games while being aware of the ways in which one is being played by them, thereby heightening the gamer's critical distance and improving his reflective skills." (Kringiel, p. 639). *Game use literacy* involves the interactional competencies needed to play and move forward in the game. Finally, *game production literacy* is developed

---

[2] This claim is based on a study by the Pew Research Center from 2003, in which students at 27 institutions of higher education across the US participated (Jones, 2003). This number is all the more striking when compared available data on the literary reading practices of Americans. For example, *Reading at Risk*, a report by the National Endowment for the Humanities (Bradshaw & Nichols, 2004), found that only 66.7% of Americans with a college degree reported reading a literary work outside of a school context. These numbers were even lower among other demographics.

[3] In his developmental theories, Vygotsky posits that there is no play without rules. "One could go even further and propose that there is no such thing as play without rules. The imaginary situation of any form of play already contains rules of behavior, although it may not be a game with formulated rules laid down in advance .... What passes unnoticed by the child in real life becomes a rule of behavior in play" (Vygotsky, 1978, p. 93).

through an awareness of the many factors and choices made in game design and production. Holistic approaches to game literacy development, which address all four of these components, thus emphasize not only playing games, but also studying and designing games from various angles and perspectives.

Echoing a distinction made by new literacy scholars between events and practices (e.g., Street, 1988), Squire (2008b; see also 2003, 2008a) introduced another productive distinction for conceptualizations of game literacy: a distinction between games as artifacts and game playing as a social practice. As artifacts, games can be seen as narrative spaces in which action is dynamic and transformative. As social practices, they afford participation in at times diffuse, global communities in which, to paraphrase Kramsch and Whiteside's (2008) discussion of language ecology in multilingual settings, language users have to navigate unpredictable exchanges and exercise symbolic power in various unanticipated ways (p. 646; see also Rampton, 1995). Game literacy entails design, in the intentionally ambiguous sense described by Kern (2000) in his work on literacy and language teaching; namely, design is both product and process. The distinction between artifact and practice, or product and process, ought not to be seen as a rigid because it is often difficult to partition one concept from the other. Still, from a pedagogical point of view, the contrast provides two different ways of analyzing and working with digital gaming in an educational context as an authored set of texts, images, and scripts on the one hand and as a dynamic form of interactional practice on the other.

Zimmerman's (2007) framework for game literacies is comprised of interrelated system, play, and design literacies. In Zimmerman's notion of design, the focus is on the dynamic process, and it is closely related to the other two aspects of game literacy in his model, play and system literacy. *System literacy* denotes "understanding the world as dynamic sets of parts with complex, constantly changing interrelationships—seeing the structures that underlie our world, and comprehending how these structures function" (p. 25). Zimmerman posits that *play literacy* is dispositional, or "a ludic attitude that sees the world's structures as opportunities for playful engagement" (p. 27). Play literacy is the capacity to recognize elements of play like mimesis, competition, disorientation, and chance (Caillois, 1961) in everyday activity, and to act on the affordances they offer for innovation, creation, and transformation. Game *design literacies* thus develop through playful experimentation with, and purposeful structuring and restructuring of, game systems, not by mastering particular sets of structures or moves per se.

Like Zimmerman's model of game literacy, the model proposed by Gee emphasizes that games are complex human systems. Gee (2007) describes the development of game literacies by explaining that when people play games actively and critically, they learn to experience the world in new ways, gain access to new affinity groups, develop resources that may be applied to future situations, and learn "how to think about semiotic domains as design spaces that engage and manipulate in certain ways" (p. 38). In other words, by playing games critically, one can potentially develop the awareness that many human systems—not just games—are purposefully designed and so represent those systems from particular subject positions.

In sum, game literacies can be understood to involve awareness that games and gaming denote complex interactions between texts, practices, and experiences, underlain by dynamic, structured, and complex rule-based, human systems. From an ecological standpoint (van Lier, 2000), game literacies include the capacity to act on recognized affordances and make meaning, or participate in semiotic practices that are *gameful* (McGonigal, 2011). Applied to L2 learning, these literacies would potentially transfer metalinguistically to the awareness that language, both familiar and those under study, are complex and dynamic, yet structured, rule-based *systems* of interrelated systems, for example, lexicogrammar and discourse, shaped by context of situation and culture (Halliday, 1978; Kramsch, 1993). They would also transfer into awareness that that languages are made real through creative language use, which is afforded by *playfulness* and a ludic disposition, and the awareness that languages, discourses, and particular genres, as social practices, are systems *designed* by and for humans, as both process and product.

2.3 Genre-Based Perspectives

Although frameworks like those of Gee and Zimmerman were developed with more general education contexts in mind, the attention to linguistic and semiotic designs and emphasis on meaningful interaction already suggest multiple points of overlap between discussions within fields of game literacy and L2 literacy. Scholars in both fields of study emphasize that literacy is born in social action. Writing on game-based literacy, Squire (2005, 2008b) argues that a *functional epistemology*, which emphasizes performance as the primary organizing principle for literacy practices, characterizes gaming discourses. As evidence of this epistemology, Squire (2008b, citing Clinton, 2004) describes the typical behavior of gamers when first encountering a new game as pressing "different buttons on the game controller (e.g., console control or computer joystick), seeing what each character (or other player role) can do in the world, because it is through what we do in game worlds that we come to know them" (p. 651). We can compare this with the observation of Frances Christie, one of the pioneers of genre-based pedagogies, that genres "determine the roles taken up by the participants, and hence the kinds of texts they are required to construct" (Christie et al, 1990, p. 16). Charles Bazerman (1994), whose work has also been seminal in genre-oriented new literacy studies, similarly described genres as "forms of life, ways of being" (p. 79).

If genres, indeed all forms of meaning-making, are to be understood as ways of being in the world, one might ponder what it is that differentiates game literacy from literacy more generally defined. As we have previously noted, theoretical considerations of game literacy do share much in common with broader discussions of multiliteracies and new literacy. Within genre-based approaches specifically, a distinction is often made between three primary conceptual strands: (a) new rhetorical approaches, which have been heavily influence by poststructuralism and emphasize relations between speech genres and recurring social situations (e.g., Freedman & Medway, 1994; Coe, Lingard, & Teslenko., 2002; Bazerman, 1997; Devitt, Reiff, & Bawarshi, 2004); (b) scholarship in English for Specific Purposes (ESP), which is more linguistic in nature and focuses on the relationship between genre structures and the particular purposes of a given discourse community (e.g., Swales, 1990); and (c) work in systemic functional linguistics (SFL), which is largely based on the lexicogrammatical models of M.A.K. Halliday and which shares with ESP an interest in how language and context are systematically interconnected (e.g., Christie & Martin, 1997; Martin & Rose, 2008). As Hyland (2003) notes, these approaches share an interest in the relations between form, purpose, and situated social context, but also differ importantly in the relative emphasis that they place on each of these aspects and the frameworks through which they make sense of them, and consequently the scholarly and pedagogical implications drawn from them diverge significantly (pp. 21-22).

In many ways, ESP and SFL have had the greatest impact on genre-based L2 pedagogy, perhaps because of their more explicit attention to linguistic features like register, style, lexis, and genre moves (functional principles within a given text). The primary educational objective of L2 teaching is paired with a social imperative, namely to provide learners with effective access to the genres of power (see Hyland, 2003; NLG, 1996). The renewed enthusiasm for research on advanced language learning over the past decade or so has been closely affiliated with these two areas of genre study (e.g., Byrnes & Maxim, 2004; Byrnes, Maxim, & Norris, 2010). Discussions around what has been described as the bifurcation of many foreign language programs, the division between lower-level language programs geared towards spoken communication and upper-level courses featuring content-based courses in literature and culture, have also tended to posit genre as an organizing principle for articulating curricular sequences (e.g., Swaffar & Arens, 2005). Such curricula are characterized by a move from more personal, concrete realities and meanings and texts that are linguistically less complex to a focused engagement with extended discourse, including the negotiation and articulation of alternate or abstract realities, complex and ambiguous meanings, and supersentential, discourse-level processing, or what Maxim (2006) terms *textual thinking*.

2.4 Integrating Genre and Game Literacies

The curricular context for the unit on gaming reported here was structured according to the kind of generic progression described above. The 16-week, 6-credit-hour German language course was largely organized around three writing tasks: a descriptive text, a narrative, and an argumentative essay. Each task involved three drafts, each requiring students' attention to purpose and genre, with an increasing attention to lexis and style starting in the second and grammatical accuracy only in the third drafts. Each of the main thematic units (see Table 1), included a variety of texts belonging to or sharing features with the genre family invoked by the writing task.[4]

Table 1
Curriculum Overview

Special Topics	Featured Genre Family
Das Individuum in der Gesellschaft 'The Individual in Society'	Description/Portrayal/Vignette
Geschichten aus der deutschen Geschichte 'Stories from German History'	Narrative
Gaming und Spielkulturen 'Gaming and Game Culture'	Argumentative essay

This type of genre-based multiliteracies curriculum has been championed by scholars within foreign language pedagogy and language teaching as an effective means for promoting learners' access to and awareness of textual practices and the lexicogrammatical features associated with them (e.g., Byrnes, 2009; Ryshina-Pankowa & Liamkina, 2012). The *Design* metaphor introduced by the NLG (1996) and developed within L2 teaching and learning by Kern (2000), among others, is realized in the curriculum through a careful engagement with texts, and in particular literary texts. Because literary works often provoke, parody, and parrot other genres, they are potentially productive for heightening students' awareness of how meaning-making occurs within systems of choices; however, the heavy emphasis on text-based pedagogies within foreign language teaching and learning, which have been heavily shaped by Hallidayan systemic functional paradigms, has arguably left nonlinguistic aspects of genre neglected. The focus on genre parameters that can be meticulously described using linguistic models has led to an emphasis on interpretation and critical language awareness, both of which are arguably highly desirable; however, all too often, learners are positioned as peripheral and even detached participants within these frameworks (see Warner, in press). In such approaches, attention to the participatory and situated nature of language use (see NLG, 1996; Bawarshi, 2000) is relegated to context.

Whereas design, interpretation, and critical reflection describe aspects of a language user's ability to apprehend and produce complex utterances in situated, sociocultural contexts, the scholarship on gaming literacy asks us also to account for the precarious, socioaffective experience of semiotic activities. Within gaming studies, this tension is witnessed in the so-called narratology/ludology debates (Aarseth, 1997; Juul, 2005; Murray, 2013), which revolve around the extent to

---

[4] These units were preceded by a short introductory unit, which emphasized reading strategies, and a final unit in which students read Wladimir Kaminer's *Mein deutsches Dschungelbuch*, a collection of travel narratives from around Germany that combined many aspects of the three genre families students had already encountered and functioned as a kind of course capstone.

which games can be legitimately analyzed as representational, narrative texts or whether they should be analyzed primarily as rule-based, playful systems, that is, as a particular form of psychosocial activity. For the purposes of this study, an intermediary position was taken, which recognizes both perspectives as legitimate and furthermore reframes the questions phenomenologically in terms of how a given user regards gameful interaction in a particular moment (see also Squire, 2008b). By integrating gaming into the existing genre-based multiliteracies curriculum, we hoped also to bring more of the social life of genres, the extent to which speech genres are ways of being in the world, into the classroom and with some of the precarity and contingency that perhaps get lost in exclusively linguistic models of genre pedagogy.

3. An Instructional Unit for Digital Games as Practices and Texts

Within the existing curriculum, the inclusion of a unit on digital gaming was seen as an opportunity to enable students to reflect on language use and gameful semiosis through an oscillation between play and the critical framing of play, thus developing L2-mediated game literacies. For this reason, in-class and outside-of-class activities during the 2-week module fell into two mutually enriching but quite distinct categories: (a) actual gameplay activity and (b) reading and writing about gaming discourses.

The instructors provided students with the URLs of two German browser game companies, Innogames and Gameforge, from which students could choose the games to play at home for the two weeks of the unit. We chose browser games because they provided several practical advantages. First, since no download is necessary, the games are accessible on various digital devices, including those found in campus computer labs, which often prohibit software installation. Second, browser games are typically free and only require a valid email address for registration. Additionally, the two sites offered students a number of different titles, which we felt would complement preferred game genres, playing styles, and personal interests.

Students began with a questionnaire about gaming habits and preferences, and took the Bartle Test of Gamer Psychology (1996), a casual quiz popular among gamers to identify their gameplay style as achiever, killer, socializer, or explorer. Students were then instructed to explore the websites and try out the different games individually, after which they were put into groups and directed to choose one game to play over the following 2 weeks. In this way, student interactions around gameplay would afford the development of learner-player community. The four groups of students agreed upon three games: Wild Guns, Ikariam, and Forge of Empires, all single-player casual strategy games in which the objective is to design, build, and conquer a place, whether it be an empire, a settlement, or a city. The games are quite similar, and diverge mostly in their settings—the 19th century American frontier, Roman antiquity, or Western European history. Students were required to play their game four times a week for at least 20 minutes per session and then to reflect on their gameplay in a private log in the course management system. In their logs, students were to note new words and phrases and to reflect on what they were doing in the game, what they felt they were learning, and to what extent they enjoyed the experience.

Aligning with the genre-based, cultural studies approach of the curriculum, students were also asked to read texts or view videos that thematized current discourses about digital gaming practices, especially in the German-speaking world. The topics included discussions of game genres and gameplay styles, gaming communities, gaming and identity, sexism and violence in games, and gaming and learning. Some of the activities also involved participation in digital literacy practices, such as creating avatars with bitstrip.com, searching and posting in gamer forums, and chatting in Google Docs and in one of the game's global chat. The discussion topics also brought the learners' chosen games into the classroom. Topics such as game sexism and violence and game forum discourses were connected to the examples from their chosen games.

4. Results: Student-Player Tensions

The game log entries as well as spontaneous feedback and semester reflections show that the gaming unit received mixed reactions and led to unexpected learning. Some students embraced games in the classroom, enjoyed playing their game, and experienced positive emotions while playing, as illustrated by the following translated excerpts from game log entries:

> Today I did not play 20 minutes. I have played about 2 hours in the morning and then 1 hour later. Construction takes a little longer than I like, but the game is still fun. I have improved my town hall, and I have researched a lot. After I have improved my academy I can start building a shipyard and get ships ready to go to war!
> Student 1, Nov 9th (translated)

> So, I founded a colony. It is ready now and I can start building something. My new city is called *Städtchen*, that's really a German word! I have also built a carpentry next to my pub. I'm adding to my town hall now and one of my ships brings goods to my *Städtchen*.
> Student 1, Nov 15th (translated)

While this student's game play seems to represent the unification of pleasure and learning cited by scholars as a key argument for gaming in education (e.g., Gee, 2007; Thorne, Black, & Sykes, 2009), her experience was not shared by all participants in the class. One of the most frequent points of dissatisfaction related to a conflict between playing and learning. For some, writing the log entries made playing feel like a chore, and others pointed out that it was strange to have homework involving play. Others did not take the homework seriously and openly admitted that they procrastinated playing in favor of "real" homework.

In contrast to those who were engaged and found pleasure in the game assignments, some students expressed frustration:

> Today the same things happened. I know that I have to fight somehow in this game, but I don't know how to do that. My people are not happy. I have not learned anything new. The game is still not fun. It's boring.
>
> Student 2, Nov 8th (translated)

> Today I fought for the first time. Somehow I won, but I'm not sure how. I especially noticed that you don't have much control in this game. It's still not fun. I don't understand what the goal of the game is. It's frustrating.
>
> Student 2, Nov 12th (translated)

Although the student advanced in the game as well (she learned how to fight and conquer), the game design frustrated her. She felt the game design gave her no agency, and it was not clear to her what she was supposed to be learning, which may indicate her frustration was in the particular genre her group selected. Another student stated specifically that had she played a game genre that she liked more, she would have probably been so into it that she would have forgotten all about learning vocabulary:

> It was a strategy game, and I like killer games. At the same time, if I had had a killer game, I think I wouldn't have learned as much, because I would have wanted to play more than to learn vocabulary.
>
> Student 3, Nov 12th (translated)

The comments of Students 2 and 3 illustrate a sense that playing and learning are two oppositional activities that may conflict, a clear challenge for game-enhanced pedagogy where "children will interpret and use videogames from outside the educational context, potentially viewing them as pure play, and thus may have difficulty in viewing them as educational instruments" (Lacasa, Martínez, & Méndez, 2008, p. 87). Digital games and gaming culture are understood as vernacular and counterhegemonic to dominant educational discourses where learning is equated with schoolwork, coursework, and homework. A fourth student—a self-proclaimed nongamer—developed his own strategy to complete the assignment by designing a computer program to play the game for him. This strategy landed him in trouble:

> After this last reflection this morning, I have just tried to play again. When I register, I encountered a ticker. The game suspends me for 24 hours. The company 'Inno Games' cites breach of the official rules. They talk specifically about the harms of 'bot' and 'script-use' ... I feel very happy and when I play the game in the future I will play fair. My 'achiever' qualities have me in trouble.
>
> Student 4, Nov 10th (translated)

Because of his ban, Student 4 "learned more about legal German than ever before" and found himself faced with genres that neither he nor the instructors had anticipated. It is ironic—but also noteworthy—that the student's de Certeauean

(1984) strategy, his attempt to "game" the assignment landed him in this unexpected situation. It is also noteworthy that the student, who clearly established his identity as a nongamer at the beginning of the unit, recast his identity as an *achiever* who perhaps found his assigned game genre unappealing. Although instructors may not envision or desire such an instructional outcome, it is exactly such gameful, unpredictable opportunities that afford the development of literacies and the experience of genres as social action.

5. Discussion and Implications

Our pilot implementation revealed that integrating digital games into a genre-based, advanced L2 curriculum proved challenging, primarily because of tensions between play and learning. While about half our students enjoyed the gaming unit and found it effective for learning, the other half—including some avid gamers—did not. Several points of student criticism related to the relationship between the games and the rest of the course curriculum. In spite of the fact that the work with texts on topical issues related to gaming was similar to what these students had done during the rest of the semester, they commented that the gaming unit lacked connection with the rest of the course and that the implementation focused too much on assignments and learning. This criticism is similar to Lacasa et al.'s (2008) assertion that learners may have trouble viewing games as educational objects. A tension seems to arise from this phenomenological attitude towards games as "pure play" (p. 87). In asking students to document the experience of gaming in a game log, the instructors hoped to facilitate reflection, but this additional assignment coupled with the course requirement to play also made gaming unnatural and more chore than play. Writing about a fun activity, like digital gaming, in a foreign language is a challenging task, and for some students, doing so negatively impacted the actual gaming experience. For this reason, the gaming unit was—at least for some students—both too educational (and consequently, not enough fun) and not educational enough because the object was perceived as nonserious and distinct from the other forms of less vernacular, "higher" cultural production (primarily literary texts and films) which constituted most of the remainder of the curriculum.

Resolving this tension is certainly difficult and may not even be wholly desirable; these perceptions are as much a part of the cultural practice of gaming as the other aspects of meaning-making, which were highlighted in the unit. Nevertheless, we can make several suggestions that might alleviate the work of keeping a reflective log in a foreign language while still allowing students to document their gaming process. For example, students could get more time to play in class, allowing their process to be monitored and to build a sense of player community. Students could be assigned game partners with whom they could chat in their game, or if the game did not have a chat function, the students could use an external chat tool. Integrating games more effectively might also involve foregrounding features that the games share in common with the literacy practices featured in the other units of the course. For instance, if students play a game that allows for the creation of characters/avatars, students could write their first essay—the description—about one of those figures. For games with simulation features, students could write stories related to their gameworld, in text or multimodal format.

Much of the negative feedback related to the characteristics of the games themselves. Although browser games offer numerous advantages, like access, cost, and variety of themes, many students did not enjoy the ones they selected, primarily because they were all of the same genre, even though they were topically dissimilar. Responsive activities might expand on the Bartle test and lead to discussion of how player style interacts with game genre and how, like genres of any media type, game genres are hybrid and have evolved through social action and purposeful design. While games may seem topically dissimilar, like any other text, they may be similar in genre to others because of common structure or social purpose, and as texts, games embody and reflect discourses. Activities might have as their objective the development of system, play, and design literacies through the analysis and experience of games as texts and gaming as practice. For the L2 classroom, these literacies would ideally transfer to awareness of the systemic, playful, and designed qualities of language itself.

6. Conclusion

One of the key proponents of New Rhetorical approaches, Charles Bazerman (1997), describes genres as "the familiar places we go to create intelligible communicative action with each other and the guideposts we use to explore the familiar" (p. 19). Foreign language learning is, more or less by definition, the exploration of the unfamiliar and the new. Using a comparable spatial metaphor—but one which may feel more applicable to the L2 teaching and learning context—Jenkins and Squire (2004) theorize video games as an art of *contested spaces*. Gameworlds are contested because users must position themselves in territories that are deliberately designed for "creating conflict within space" (Squire, 2008a, p. 646) or tension between player agency and game design—the back-and-forth of personal and designed narrative (Neitzel, 2005) that emerges through play. Although games are designed to provoke certain actions by calling upon users' gaming literacies, the best games are those that afford a sense of agency and creative, critical engagement. Not in spite of, but because of all of the unpredictable interactions and situations in which learners might find themselves when playing in a foreign language, games might allow learners to explore communicative and symbolic practices as a negotiation between the familiar and the contested.

If we accept this exploration of familiar and contested social spaces as an objective of L2 learning, new questions about potential roles of digital gaming in L2 teaching and learning surface. In the curricular context of the course we describe here, awareness-raising activities related to various lexicogrammatical features of genres provided textual guideposts by which students could become more familiar with various forms of expression. The 2-week gaming unit attempted to capture some of the complexity of expression as social action and, in effect, was transformative in predictable and unpredictable ways. While we can plan for contingencies in future implementations, we also expect the unexpected and hope to leverage unforeseen complexities into learning opportunities.

We conclude here with a final question for future investigations of digital gaming in instructed L2 contexts: What might it mean to take games—not just textuality—as an organizing principle for foreign language curricula? Future studies might engage with this question by considering how "gamefulness" (McGonigal, 2011; Reinhardt & Thorne, in press) might develop learners' sense of how to use the meaning-making resources emphasized in text-oriented pedagogies to "game" the system, twisting and bending rules in order to design social futures (NLG, 1996) in new languages. In other words, we might recognize that the literacies developed through critical gaming—systems thinking, design, and playfulness—may have applicability to language development more broadly. As Zimmerman has stated emphatically, "gaming literacy is not about just about playing games" (2008, p. 30).

References

Aarseth, E. (1997). *Cybertext: Perspectives on ergodic literature*. Baltimore, MD: Johns Hopkins University Press.

Baacke, D. (2004). Medienkompetenz als zentrales Operationsfeld von Projekten. In S. Bergmann, J. Lauffer, L. Mikos, G. Thiele, & D. Wiedemann, Dieter (Eds.), *Medienkompetenz: Modelle und Projekte* (pp. 21–25). Bonn, Germany: Bundeszentrale für politische Bildung.

Bartle, R. (1996). *Hearts, clubs, diamonds, spades: Players who suit MUDs*. Retrieved from http://www.mud.co.uk/richard/hcds.htm

Barton, D. (2007). *Literacy: An introduction to the ecology of written language*. London, England: Blackwell.

Bawarshi, A. (2000). The genre function. *College English, 62,* 335–360.

Bazerman, C. (1994). Systems of genres and the enactment of social intentions. In A. Freedman & P. Medway (Eds.), *Genre and the new rhetoric* (pp. 79–101). London: Taylor and Francis.

Bazerman, C. (1997). The life of genre, the life in the classroom. In W. Bishop & H. Ostrum (Eds.), *Genre and writing* (p. 19–26). Portsmouth, NH: Boynton/Cook.

Bolitho, R., Carter, R., Huges, R., Ivanic, R., Masuhara, H., & Tomlinson, B. (2003). Ten questions about language awareness. *ELT Journal, 57,* 251–259.

Bradshaw, T., & Nichols, B. (2004). *Reading at risk: A survey of literary reading in America*. Washington, DC: National Endowment for the Arts. Retrieved from http://arts.gov/sites/default/files/ReadingAtRisk.pdf

Buckingham, D. (2003). *Media education: Literacy, learning, and contemporary culture*. London, England: Blackwell.

Byrnes, H. (2009). Emergent L2 German writing ability in a curricular context: A longitudinal study of grammatical metaphor. *Linguistics and Education, 20*, 50–66.

Byrnes, H., & Maxim, H. (Eds.). (2004). *Advanced foreign language learning: A challenge to college programs*. Boston, MA: Thomson Heinle.

Byrnes, H., Maxim, H., & Norris, J. (2010). Realizing advanced foreign language writing development in collegiate education: Curricular design, pedagogy, assessment [Monograph]. *Modern Language Journal, 94*(s1).

Caillois, R. (1961). *Man, play, and games*. Glencoe, IL: The Free Press.

Christie, F., & Martin, J. (Eds.). (1997). *Genre and institutions: Social processes in the workplace and school*. London, England: Cassell.

Christie, F., Gray, P., Gray, B., Macken, M., Martin, J. R. and Rothery, J. (1990) *Language a Resource for Meaning: Procedures. Reports. Explanations*. Sydney, Australia: Harcourt Brace Jovanovich.

Coe, R., Lingard, L., & Teslenko, T. (2002). *The rhetoric and ideology of genre*. Cresskill, NJ: Hampton Press.

Cook, G. (2000). *Language play, language learning*. Oxford, England: Oxford University Press.

Cope, B., & Kalantzis, M. (2000). *Multiliteracies: Literacy learning and the design of social futures*. New York, NY: Routledge.

Cornillie, F., Thorne, S., & Desmet, P. (2012). Digital games for language learning: From hype to insight? Digital games for language learning: Challenges and opportunities [Special issue]. *ReCALL, 24*, 243–256.

de Certeau, M. (1984). *The practice of everyday life*. Berkeley, CA: University of California Press.

Devitt, A., Reiff, M., & Bawarshi, A. (2004). *Scenes of writing: Strategies for composing with genres*. New York, NY: Longman.

Freedman, A., & Medway, P. (1994). *Genre and the new rhetoric*. London, England: Taylor & Francis.

Gee, J. (1991). What is Literacy? In C. Mitchell & K. Weiler (Eds.), *Rewriting literacy: Culture and the discourse of the Other* (pp. 3–11). New York, NY: Bergin & Garvey.

Gee, J. (1992). *The social mind: Language, ideology, and social practice*. New York, NY: Bergin & Garvey.

Gee, J. (1996). *Social linguistics and literacies: Ideology in discourses*. London, England: Falmer Press.

Gee, J. (2003). *What video games have to teach us about learning and literacy*. New York, NY: Palgrave Macmillan.

Gee, J. (2004). *Situated language and learning: A critique of traditional schooling*. New York, NY: Routledge.

Gee, J. (2007). Pleasure, learning, video games, and life: The projective stance. In M. Knobel & C. Lankshear (Eds.), *A new literacies sampler* (pp. 95–113). New York, NY: Peter Lang.

Halliday, M. (1978). *Language as social semiotic: The social interpretation of language and meaning*. London, England: Arnold.

Hyland, K. (2001). *Genre and second language writing*. Ann Arbor: University of Michigan Press.

Hyland, K. (2003). Genre-based pedagogies: A social response to process. *Journal of Second Language Writing, 12*, 17–29.

Jenkins, H., & Squire, K. (2004). Harnessing the power of games in education. *InSight, 3*, 5–33. Retrieved from http://sites.edvantia.org/products/pdf/InSight_3-1_Vision.pdf

Jenkins, H., Purushotma, R., Clinton, K., Weigel, M., & Robinson, A. (2006). *Confronting the challenges of participatory culture: Media education for the 21st century*. Chicago: MacArthur Foundation. Retrieved from http://www.macfound.org/media/article_pdfs/JENKINS_WHITE_PAPER.PDF

Jones, S. (2003). Let the games begin: Gaming technology and college students. *Pew Research Internet Project*. Retrieved from http://www.pewinternet.org/2003/07/06/let-the-games-begin-gaming-technology-and-college-students/

Juul, J. (2005). *Half-real: Video games between real rules and fictional worlds*. Cambridge, MA: MIT Press.

Kern, R. (2000). *Literacy and language teaching*. Oxford, England: Oxford University Press.

Kramsch, C., & Whiteside, A. (2008). Language ecology in multilingual settings: Towards a theory of symbolic competence. *Applied Linguistics, 29*, 645–671.

Kramsch, C. (1993). *Context and culture in language teaching*. Oxford, England: Oxford University Press.

Kress, G. (2003). *Literacy in the new media age*. London, England: Routledge.

Kringiel, D. (2012). Learning to play: Video game literacy in the classroom. In J. Fromme & A. Unger (Eds.), *Computer games and new media cultures: A handbook of digital games studies* (pp. 633–646). Berlin, Germany: Springer.

Lacasa, P., Martínez, R., & Méndez, L. (2008). Developing new literacies using commercial videogames as educational tools. *Linguistics and Education, 19*, 85–106.

Lankshear, C., & Knobel, M. (2006). *New literacies: Everyday practices and classroom learning.* New York, NY: Open University Press.

Martin, J., & Rose, D. (2008). *Genre relations: Mapping culture.* London, England: Equinox.

Maxim, H. (2006). Integrating textual thinking into the introductory college-level foreign language classroom. *Modern Language Journal, 90,* 19–32.

McGonigal, J. (2011). *Reality is broken: Why games make us better and how they can change the world.* New York, NY: Penguin.

Murray, J. (2013, June 28). The last word on ludology v narratology (2005) [Web log message]. Retrieved from http://inventingthemedium.com/2013/06/28/the-last-word-on-ludology-v-narratology-2005/

National Council of Teachers of English (2013). *NCTE framework for 21st century curriculum and assessment.* Urbana, IL: National Council of Teachers of English. Retrieved from http://www.ncte.org/library/NCTEFiles/Resources/Positions/Framework_21stCent_Curr_Assessment.pdf

Neitzel, B. (2005). Narrativity of computer games. In J. Goldstein & J. Raessens (Eds.), *Handbook of computer game studies* (pp. 227–245). Cambridge, MA: MIT Press.

New London Group. (1996). A pedagogy of multiliteracies: Designing social futures. *Harvard Educational Review, 66,* 60–92.

Phillips, M. (1987). Potential paradigms and possible problems for CALL. *System, 15,* 275–287.

Rampton, B. (1995). *Crossing: Language and ethnicity among adolescents.* London, England: Longman.

Reinders, H. (Ed.). (2012). *Digital games in language teaching and learning.* New York, NY: Palgrave Macmillan.

Reinhardt, J., & Thorne, S. (2011). Beyond comparisons: Frameworks for developing digital L2 literacies. In N. Arnold & L. Ducate (Eds.), *Present and future promises of call: From theory and research to new directions in language teaching* (pp. 257–280). San Marcos, TX: CALICO.

Reinhardt, J., & Thorne, S. (in press). Teaching languages with digital games. In F. Farr & L. Murray (Eds.), *Routledge handbook of language learning and technology.*

Ryshina-Pankowa, M., & Liamkina, O. (2012). Grammar dilemma: Teaching grammar as a resource for making meaning. *Modern Language Journal, 96,* 270–289.

Squire, K. (2003). Video games in education. *International Journal of Intelligent Simulations and Gaming, 2, 49–62.*

Squire, K. (2005). Toward a media literacy for games. *Telemedium, 52(1-2),* 9–15. Retrieved from http://website.education.wisc.edu/kdsquire/tenure-files/24-telemedium.pdf

Squire, K. (2008a). Open-ended video games: A model for developing learning for the interactive age. In K. Salen (Ed.), *The ecology of games: Connecting youth, games, and learning* (pp. 167–198). Cambridge, MA: MIT Press.

Squire, K. (2008b). Video game literacy: A literacy of expertise. In J. Coiro, M. Knobel, C. Lankshear, & D. Leu (Eds.), *Handbook of research on new literacies* (pp. 635–670). Mahwah, NJ: Erlbaum.

Steinkuehler, C. (2005). *Cognition and learning in massively multiplayer online games: A critical approach* (Unpublished doctoral dissertation). University of Wisconsin, Madison, WI.

Street, B. (1984). *Literacy in theory and practice.* Cambridge, MA: Harvard University Press.

Street, B. (1988). Literacy practices and literacy myths. In R. Saljo (Ed.), *The written word: Studies in literate thought and action* (pp. 59–72). Heidelberg, West Germany: Springer.

Street, B. (1995). *Social literacies: Critical approaches to literacy in development, ethnography and education.* New York, NY: Longman.

Street, B. (2003). What's "new" in New Literacy Studies? Critical approaches to literacy in theory and practice. *Current Issues in Comparative Education, 5,* 77–91.

Swaffar, J., & Arens, K. (2005). *Remapping the foreign language curriculum: An approach through multiple literacies.* New York, NY: Modern Language Association.

Swales, J. (1990). *English in academic and research settings.* Cambridge, England: Cambridge University Press.

Sykes, J., & Reinhardt, J. (2012). *Language at play: Digital games in second and foreign language teaching and learning.* New York, NY: Pearson.

Sykes, J., Reinhardt, J., & Thorne, S. (2010). Multi-user gaming as sites for research and practice. In F. Hult (Ed.), *Directions and prospects for educational linguistics* (pp. 117–135). Dordrecht, Netherlands: Springer.

New London Group. (1996). A pedagogy of multiliteracies: Designing social futures. *Harvard Educational Review, 66,* 60–92.

Thorne, S., & Reinhardt, J. (2008). "Bridging activities," new media literacies, and advanced foreign language proficiency. *CALICO Journal, 25,* 558–572.

Thorne, S., Black, R., & Sykes, J. (2009). Second language use, socialization, and learning in internet interest communities and online gaming. *Modern Language Journal, 93,* 802–821.

van Lier, L. (2000). From input to affordance: Social interactive learning from an ecological perspective. In J. P. Lantolf (Ed.), *Sociocultural theory and second language learning* (pp. 245–259). Oxford, England: Oxford University Press.

Vygotsky, L. (1978). *Mind in Society*. Cambridge, MA: Harvard University Press.

Warner, C. (in press). Mapping new classrooms in literacy-oriented foreign language teaching and learning: The role of the reading experience. In J. Swaffar, P. Urlaub, & K. Arens (Eds.), *Transforming the foreign language curriculum in higher education*.

Zimmerman, E. (2007). Gaming literacy – game design as a model for literacy in the 21st century. Retrieved from http://ericzimmerman.com/files/texts/Chap_1_Zimmerman.pdf

Zimmerman, E. (2008). *Gaming literacy: Game design as a model for literacy in the 21st century*. Retrieved from: http://llk.media.mit.edu/courses/readings/Zimmerman-Gaming-Literacy.pdf

Chapter 8

Integrating Digital Stories in the Writing Class: Toward a 21st-Century Literacy

ANA OSKOZ
University of Maryland, Baltimore County (USA)

IDOIA ELOLA
Texas Tech University (USA)

Abstract

> This chapter presents the integration of digital stories (DSs)—individually created online storylines that require the successful integration of text, images, and sounds—by six learners in an advanced Spanish writing class to enhance learners' 21st-century digital skills. Following a task-based approach, learners developed a DS through topic investigation, revision of narrative drafts, and inclusion of images and sound. Additionally, learners participated in (a) 11 online journals designed to analyze the learners' approaches to the narrative text and its development into an oral text, and to the integration of image and sound; (b) a final reflection regarding the DS development process; and (c) a questionnaire designed to clarify learners' DS working process and its benefits and drawbacks.
>
> Analyses of the narrative drafts illustrate learners' choices when integrating content from academic assignments into their own DS. In addition, analyses of the online journals, questionnaires, and reflection pieces suggest that, despite the initial difficulties encountered when developing a DS, learners followed the thematic and writing conventions of the course's academic assignments. Creating a DS also pushed learners to move beyond traditional oral and written presentational forms by making connections to other forms of expression (i.e., images and sound).

1. Introduction

Writing in a foreign language (FL) has been recently much studied in terms of the integration of new social tools such as wikis, blogs, Facebook, and Twitter in the classroom. These media have helped students to use different registers in their writing; for example, recent studies have focused on the use of wikis for formal academic writing or blogs for more informal personal writing (Elola & Oskoz, 2011) and examined the writing process or final text product (Blin & Appel, 2011; Kessler, 2009). However, the oral/story-telling potential of written texts has not been similarly explored. Digital stories (DSs)—online storylines that require the integration of text, images, and sounds—have proved an ideal vehicle to explore the oral aspect of texts. DSs have already been widely employed in education (Lambert, 2002, Robin, 2006), notably in ESL (Rance-Roney, 2008; Vinogradova, 2008; Vinogradova, Linville, & Bickel, 2011) and in the context of heritage speakers (Vinogradova, 2014). Yet less is known about their potential in the FL classroom (for an exception, see Castañeda & Rojas-Miesse, 2013). Either in the form of personal narratives or as background to course-related topics, DSs can empower learners by allowing them to share their voices and their views in an open and interactive environment (Gregori-Signes, 2008; Robin, 2008; Sadik, 2008). The development of a DS, however, involves more than simple inclusion of images, music, and voice. It is the integration and combination of these three elements in complex layers that promotes learners' linguistic and writing development. The integration of a DS in a writing class might also challenge learners in new ways because this conception of the writing act shifts away from traditional pen-and-paper or word-processing methods into areas that initially might seem novel and daunting.

In this study we explore how learners approach the process of writing with tools such as iMovie. Following a task-based approach (Samuda & Bygate, 2008), learners developed a DS product through the use of collected images, topic investigation, revision of drafts, and final product. They worked individually and in collaboration with their classmates and the instructor; the use of "circles" to share feedback was incorporated. Based on Activity Theory (Leontiev, 1978), this study examines what happens when learners are confronted with a new type of writing genre that challenges traditional assumptions of the writing act. In particular, we examine the impact that new artifacts, such as iMovie or Final Cut, have on the object (the DS product); how the activity is orientated from an initial object (the digital story) towards an eventual outcome for the learners themselves (become effective 21st-century writers); how the introduction of new rules (multimedia inclusions) influences the division of tasks in the FL classroom; and how learners come to terms with a new genre.

The study took place in an advanced Spanish writing course. The participants, after working with traditional writing assignments, were asked to create a DS based on one of the topics discussed in class. This study investigates (a) what the linguistic and writing structural differences are between traditional forms of writing and 21st-century literacies (digital stories), (b) how learners presented the content in written and oral modalities, and (c) what the learners perceived as benefits or drawbacks following participation in the new assignment.

2. Literature Review

2.1 Task-Based Approach

Tasks have been defined as activities or actions that are carried out as a result of processing and understanding language (Richards, Platt, & Webber, 1985); that is, a goal-oriented activity that requires learners to use language to produce a tangible outcome (Prabhu, 1987; Willis, 1996). Although the design of a task might require the use of particular linguistic resources or forms (Ellis, 2003), all definitions of a task state or imply that learners should give primary attention to meaning (Skehan, 1998). From the many definitions of task, we have adopted Samuda and Bygate's (2008) definition of a pedagogical task for the present study; that is: a "holistic activity which engages language use in order to achieve some nonlinguistic outcome while meeting a linguistic challenge, with the overall aim of promoting language learning through a process or product or both" (p. 69). Two points are important in this definition. First, while in any other context language would be the means to achieve a nonlinguistic outcome, in the context of second language (L2) learning, language is both a means and an end. Second, in working towards a task outcome, learners engage constantly in language processes as they plan and organize the work, distribute subtasks, and identify and share information as necessary. This subscribes to a philosophy of learning by doing; that is, undertaking a written composition pedagogically engineered into a sequence of interconnected tasks that make the process more accessible and clear. Studies have indicated that process writing is an appropriate methodology for L2 writing because it imitates the kind of writing behavior seen in expert writers—writing that leads to successful outcomes that can be shared or published (Flower & Hayes, 1981).

By definition, a task has phases that may be broken down into several interrelated steps (Samuda & Bygate, 2008, p. 14) to make the overall task more manageable. As Samuda and Bygate state "the different phases of a task can serve different functions, and thus might give rise to different types of exchange and different types of talk" (p. 15). Analysis of individual phases can illuminate the language that students use, and how the task is viewed. In line with this approach, the writing task was therefore divided into sequenced stages which included interrelated but distinct subtasks that together formed a full sequence.

2.2 Task-Based Approach Within an Activity Theory Framework

Sociocultural perspectives have recently validated the inclusion of task-based approaches in the classroom, and they have offered some insights to steer the direction of research (Ellis, 2003). A central tenet of sociocultural theory is that higher forms of mental activity, such as attending, predicting, planning, monitoring, and inferencing, are mediated mental activities whose sources are external to the individual, but in which the learner participates through dialogue. The mediation of the cognitive functions occurs via psychological or semiotic tools, such as numbers, symbols or language, and physical tools or artifacts, such as tasks and technology (Ellis, 2003). Such tools and artifacts are not neutral and therefore their choice and use can influence language development (Thorne, 2003; Hampel & Hauck, 2006) and shape the mediation process in important ways (Werstch, 1991).

From an Activity Theory perspective, Blin and Appel (2011) highlighted the essential role of artifacts used and created by L2 learners in mediating their collaborative writing practice in online, asynchronous, moment-by-moment interactions. Activity Theory, or cultural-historical activity, is based on the premise that cognitive development has a cultural and social origin (Lantolf & Thorne, 2006). Drawing heavily on the concept of mediation—the mechanism through which external sociocultural activities are transformed into internal mental functioning (Vygostky, 1978)—Activity Theory has evolved from its origins to fit contemporary social applications. Oskoz and Elola (2012), for example, found that learners, through their interaction either in discussion boards or chats, oriented and reoriented their jointly formed actions in relation to the object and desired outcome. Despite tools and artifacts being "integral and inseparable components of human functioning" (Engeström, 1999, p. 29), the focus of study becomes the dynamic nature of the interrelationship between the components of an activity system. As shown in Figure 1, within a collective activity system, the actions of individuals occur at the nexus of three factors: the tools and artifacts available (e.g., computers, languages, and tasks), the community and its understood rules (e.g., between instructor and learners in a classroom), and the division of labor in those community settings (e.g., instructors structuring lesson units).

Figure 1
Activity System

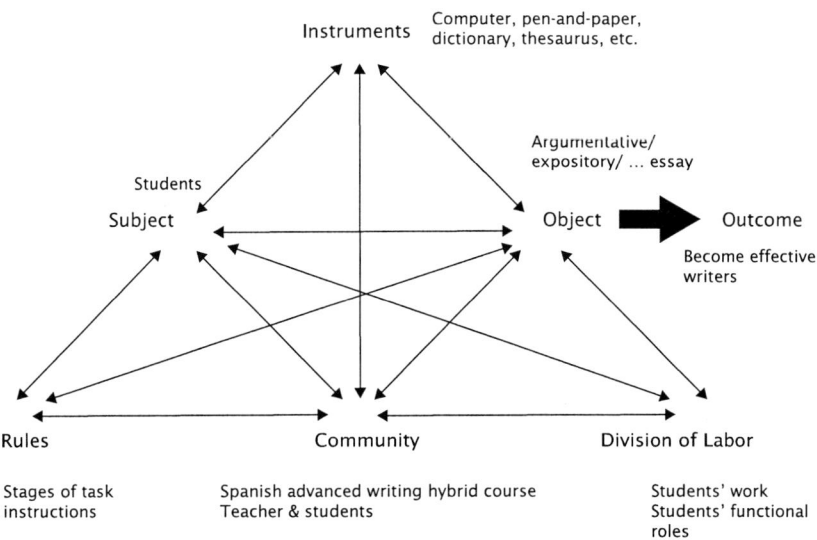

The left axis of the triangle represents the *subject*, the individual or group whose agency is, in the emic sense, the focus of the analysis (e.g., a class of learners); the instruments or *mediational tools*, which are either symbolic (e.g., the FL), or material (e.g. computers or assignments); and the *object*, or the orientation of the activity arising from the motive (e.g., essay composition), towards an eventual outcome (learning how to write in a FL). The bottom of the triangle represents the *community*, or participants (FL writer-learners) who share the same *object*; the *division of labor*, which can be either *horizontal* (actions and interactions between community members) or *vertical* (who gives feedback, who receives it); and the *regulational norms* or the *rules* that govern the division of labor within a community (e.g. when the writing activity takes place, when feedback is provided). Each of these components acts to allow or constrain transactions within a functional activity system. Therefore, tools and artifacts are just one element of an activity system, and their implementation is influenced by learners' interpretations of the object (Oskoz & Elola, 2012). The implicit and explicit rules governing the actions (e.g., schedule, required events, modes of communication), the horizontal and vertical division of labor (e.g., the constant reviewing and uploading of images to create an images bank), and the community to which learners belong (the entire class or the group members).

In the case of the written text, learners use, create, and modify a broad range of artifacts (Blin & Appel, 2011), such as task guidelines, instructor's comments, discussion boards, or digital story making tools, for their writing purposes. These textual artifacts created in the course of the L2 writing activity are also known as "material-semiotic artifacts," that is, material objects "that [carry] significant information across time and space and that serve, through local interpretation, to create coherences between distal events" (Lemke, 2000, p. 21). These artifacts link longer-term processes to shorter-term events; that is, they connect the longer-term object-oriented activity (the construction of the digital story) and shorter-term individual or collaborative goal-oriented activities (e.g., writing the text, selecting images, editing the language). Yet, at the same time, as Leontiev (1978) pointed out, learners perform an action in relation to their personal goal under differing conditions. So although all learners might have the same object, such as the completion of an essay, how they carry it out varies depending on different interpretations of the writing activity and level of motivation to see the task completed. Therefore, actions must be understood in the context in which they take place. Actions also have an operational aspect, relating to the specific conditions under which the actions take place. For example, if learners need to complete a composition, they might decide how to go about writing it (using pen and paper or a computer). Thus, actions remain the same, but the conditions in which these actions occurred can vary.

2.3 Studies on Digital Stories

Digital story telling is increasingly being used for instructional purposes (Ohler, 2008; Sadik, 2008), in ESL (Rance-Roney, 2008; Vinogradova, 2011), in heritage speakers' settings (Vinogradova, Liu, & Oskoz, 2011), and in FL education (Alcantud-Diaz, 2008, 2010; Castañeda, 2013; Gregori-Signes, 2008; Reinders, 2011). When working with the learners' first language, it has been found that DSs provide an "alternative conduit of expression for those students who struggle with writing traditional text" (Reid, Parker, & Burn, 2002). Because of the integration of text, sound, and images, DSs allow either first or second language learners to produce a multimodal artifact which strongly resembles media products they encounter during their everyday lives (Haffner & Miller, 2011). The digital story is just the tip of the iceberg (Ohler, 2008), underpinned by several processes such as choosing a topic, selecting images, and integrating sound appropriate to the story. This process, however, cannot negate the value of the written text because of the "deep language acquisition and meaningful practice" (Rance-Roney, 2008, p. 29) embedded in DSs practice. Robin (2007) emphasizes that the process of selecting a meaningful topic and writing the story about the topic are the most important elements in the digital story telling process. Therefore, without disregarding the audiovisual elements, instructors and learners need to maintain focus on the story itself (Kajder & Swenson, 2004; Ohler, 2006) and on issues of language use (Gregori-Signes, 2008). Creating their DSs allow learners to enhance their communication skills by writing creatively, organizing thoughts in coherent ways, and constructing narratives (Gakhar & Thompson, 2007; Robin, 2006). In terms of linguistic development, Reyes Torres, Pich Ponce and García Pastor (2012) found that, while errors in the use of prepositions, tenses, and possessive pronouns were still present, their first-year FL learners used more complex sentences and paid more attention to grammatical rules after the course-long digital project.

A burgeoning number of studies reflect increasing interest in the integration of the DSs in the FL classroom. Yet, little is known about how learners perceive the ways in which they familiarize themselves and learn new literacies when creating in this new genre (the digital story) and using tools such as Final Cut. Therefore, this study explores the following questions:

1. How do learners interact among components of an activity system within a collaborative, computer-mediated FL writing environment?
2. How do learners perceive linguistic and writing structural differences and similarities between traditional writing genres and digital stories?
3. What do learners see as benefits or drawbacks when developing DSs?

3. Method

3.1 Participants and Setting

This study was conducted at a midsize US university. The participants, whose ages ranged from 19 to 21, were six Spanish majors enrolled in an advanced intercultural course in writing. One of the participants was a heritage speaker, and second was a Spanish native speaker. The remaining four participants were foreign language speakers of Spanish. The class was a 3-credit 1-hour course, a capstone course of the program and mandatory for all majors in Spanish; it met one evening a week for two and a half hours. Before taking this course, learners had completed at least four courses in Spanish at the third-year level, two of which focused on the development of grammar and writing conventions.

3.2 Procedure

Among other assignments for this course, learners were asked to complete a narrative essay, two argumentative essays, two expository essays, and a DS. The narrative essay gave students practice with the writing conventions required for a digital story, such as the use of preterit and imperfect tenses, and it also introduced structural aspects such as the development of a simple narrative with a linear sequence of events or a cyclical complex set of events using flashbacks. The argumentative and expository essays also introduced content useful for the DS. For each of the latter essays, learners participated in a discussion board where they discussed the following topics: (a) a relevant character in the Latino world, (b) the relevance of the Latino vote in the US elections, (c) the Dream Act, and (d) immigration. To ensure that learners would be successful with the end product (the digital story), the development of the DS was structured in phases that took place during the entire semester (see Table 1).

Table 1
Schedule for Digital Story Development in Relation to the Writing Tasks

Phase	Schedule	Pedagogic tasks
Phase 1	Weeks 3-10	Learners and instructor briefly discussed the topic of the upcoming writing activities in class
		In the discussion board, learners: • discussed the topics of the compositions in bulletin boards • uploaded one image relevant to their topic (one image each).
Phase 2		In class, learners discussed the • topic, organization and structure of the essays monitored by teacher • images and their relation with the topic.
Phase 3		Learners • thought of topics for their final projects • searched for images (and music) to accompany their script.
Phase 4	Week 10	Learners • turned in Draft 1 of the DS • received further DS training at the language lab.
	Week 11	Online, the instructor provided comments regarding content, structure, and form.
		In class, using selected, anonymous learners' scripts, learners and the instructor discussed teacher feedback and practiced how to revise for improvement of content, structure, and accuracy.
Phase 5	Weeks 11-15	Learners continued working on their scripts (together with music and images).
Phase 6	Week 16	Learners finalized their DS, presented it to the class, and published it on the web.

The six phases provided an organizational structure for the whole process.

Phases 1-3

The objective in the first three phases was two-fold: learners, with the instructor's help, worked on their argumentative and expository essays and started selecting images for the DS. For each of the essays learners and instructor first briefly discussed the topic of the upcoming writing activity in class. After brainstorming some ideas, each group of learners decided which topic(s) to develop for the discussion board. Learners researched and discussed the topic(s), looking for ideas that supported their arguments and opinions. Then, learners and instructor conversed about the topic(s) and decided on the organization and structure of the assigned genre (argumentative or expository essay). At this point the learners decided on the topic(s) for their final projects. As they made this decision, learners were asked to find at least one image related to the content and upload it to the discussion board. The class debated the value and message of the images and discussed several of the images in terms of their level of explicitness (images reflecting literally what the narrator said) or implicitness (images that made the audience see a relationship with the narration but that were not literal descriptions). Then, the best images and accompanying music were selected for their final script. The purpose of Phase 3 was to ensure that learners settled on the topic for their DSs.

Phase 4

The learners completed the first draft of the DS narrative (5 minutes, approximately 600 words) and participated in the first DS training at the language lab. The instructor provided comments regarding content, structure, and form.

Phase 5

The learners continued working on the DSs. In addition to receiving feedback from the instructor, learners read their narratives in class to a small group of students who provided feedback regarding content, structure, and organization. Learners also discussed the feelings that the DS script and the pace of the reading evoked in listeners and the type of images to be used in the final DS.

Phase 6

The learners finalized their DS, presented it to the class, and published it on YouTube.

Apart from writing the essays and creating the DSs, learners also completed (a) 11 online journal entries recording their writing approaches for each writing assignment and image and sound selection for each discussion board topic, (b) a final reflection piece regarding their own processes when developing the DS, and (c) a questionnaire that targeted the various components as presented in the Activity Theory framework such as the use of instruments and the perceived goal of the assignment.

The researchers analyzed the online journals, the questionnaire, the reflection pieces, and the final digital story using content analysis that reflected the objectives of the research (Merriam, 1998). Content analysis revealed themes such as linguistic and rhetorical benefits of the DS, conversations about the uses of images and sounds, advantages and disadvantages of peer review (in-class circles), and approaches to working with DSs as opposed to academic writing.

4. Results

4.1 Research Question One

The first research question looked at how learners interacted among components of an activity system within a collaborative, computer-mediated FL writing environment. As seen in the questionnaires, learners became personally involved with the DS—the object—and pursued its goal from the moment they realized how the DS topic was important to them (see Figure 2).

Figure 2
Activity System for Digital Stories

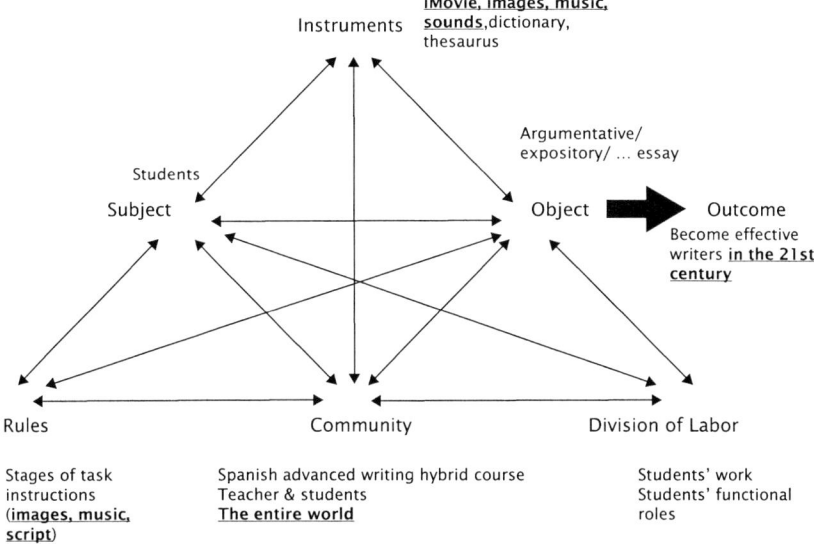

However, the questionnaires revealed that not all the participants achieved motivation in the same manner. Some became involved with the topic through the class discussions ("I decided to write on this topic after thinking about the different themes we discussed in class and how they related to me," Corine), others while writing the script ("The topic of my digital story became important to me when I started re-writing my script to make it more personal." Cora), and others through personal experiences related to some of the topics ("After spending time in a Mexican border town, I realized that the issue of immigration is seen from an entirely different perspective by non-US citizens, and I wanted to tell their side of the story," Theresa). Choosing the topic was a process in itself which in some cases took several weeks. Theresa, for example, mentioned in her journal in week five

> *Todavía estoy un poquito confundida sobre el tema de la historia digital. No sé si se puede hacer sobre cualquier asunto del cual hablamos en los tableros o sólo sobre la globalización o la inmigración. Sin embargo, ahora estoy pensado en enfocarme en los efectos negativos de la globalización (por ejemplo, el tráfico de humanos).*

'I am still a bit confused with the topic of the digital story. I don't know if we can use [any of] the topics in the discussion board or only globalization or immigration. However, now I am thinking about focusing on the negative effects of globalization (for example, human trafficking).'

Once the topic was selected, the goal of the activity started to take shape. All the learners expressed a desire to complete the DS the best way they could. They realized that they could add a personal perspective in exploring issues that touch US society. For example, Corine wanted to simply prove how feminism had impacted her life (with enough emotion to affect the audience), Soraya hoped to raise awareness of the situation that immigrant children face, and Theresa felt it was important to discuss immigration from the Mexican perspective.

Deciding on the best structure for the DS showed the different ways in which the participants approached the activity. Three main structures were chosen: (a) linear, telling a story in a chronological manner; (b) cyclical, starting and finishing with the main point of their story; and (c) flashback, moving from present to past and back. For four learners (Theresa, Corine, Soraya, and Susan), the structure was a more conscious exercise than for the other two (Cora and Matilde). For example, Theresa commented "After reworking my script draft, I realized that I wanted to make my story less linear. By starting with a 'flashback' story and then moving into a discussion of immigration, I kept my digital story more dynamic." However, Matilde's structure simply recreated the environment where she had been personally involved.

The tools and artifacts used in the digital story development were images, sounds, and obviously the language used for narration. All these elements worked together to create the digital story. Regarding the use of technology, Matilde states in her final reflection piece that

> *Mis desafíos en aprender la parte técnica fue un gran desafío para mí, porque a simple vista, me pareció muy complejo. Era como hacer mi propio manual basándome en algunas explicaciones nuevas. Cuando aprendí lo básico mi panorama cambio y se convirtió en una experiencia muy disfrutable."*

'Learning the technical part was a great challenge for me, because at first sight, it looked very complex. It was like creating my first manual based on some new explanations. Once I learned the basics, my outlook changed and it became a very enjoyable experience.'

The incorporation of sound and images as part of the story was also a novel change from their previous academic writing. Explicit images were closely related to the words describing them, whereas implicit images tended to bring out the audience's emotions. Theresa explained,

> I used images with implicit meaning more than literal images, so I think they caused the audience to think beyond the spoken words. Also, they contributed to the tone of the video. I used explicit images so that my audience could easily follow the story, but implicit images to encourage them to think beyond the spoken word.

Similarly, the integration of sound also brought a new dimension to the task. Soraya commented,

> Music provokes emotion so when you are talking about something sad then I used a song with a slow tempo and low pitch music. Then when I was talking about something that had a positive ending I used music that was more upbeat, faster-paced and higher pitched.

An interesting aspect of the use of images and sound together is their function not only as content fillers but also as connectors. Theresa stated in her journal that

> *El tono que quiero transmitir se va a intensificar a través la música. Se puede modificar el volumen para que sea más alto en los momentos más intentos y menos prominente donde la narración tiene precedencia. Al fin de la historia, quiero que cambie el tono a uno más esperanzador y optimista. Entonces, la música en esa parte será más feliz y "upbeat."*

> 'The tone that I want to transmit will be intensified through the music. The volume can be modified to make it louder in the most intense moments and less prominent where the narration itself has more prominence. At the end of the story, I want the tone to change to one which is more hopeful and optimistic. So, the music in that art will be happier and upbeat,'

Images and sounds were therefore used to move the story along as well as to connect the narration and evoke emotions.

The activity of the DS included a set of rules, to some extent unfamiliar to the learners. First, learners had to discuss the DS content in the discussion board before each of the argumentative and expository essays. Second, learners uploaded images to create a bank of images to help them with the development of their DS. Although all the phases of the DS development and information to accomplish it were provided by the instructor, working with the rules was not always easy, and it was a process the participants needed to understand and follow to fulfill their tasks. Corine affirmed,

> *Este fue una ruta complicada puesto que tenía que entender la tarea antes de empezar. Sorprendentemente, esto llevó una gran parte del semestre porque sólo pensé en mi historial digital pocos veces antes del fin del semestre. Pero, hacia el final del semestre, finalmente entendí los requisitos después de*

> hablar con mi profesora sobre cómo podía desarrollar mi tema y mi guión, para que podrían estar de acuerdo con el objetivo del proyecto.

'This was a complicated path because I needed to understand the task before I started. Surprisingly, this[feeling] lasted a big part of the semester because I only thought about my digital story a few times [...], But, towards the end of the semester, I finally understood the requirements after talking with my professor about how I could develop my topic and my script so they would meet the objective of the project.'

The adoption of the DS also influenced the perceived community of the writers. Given that a DS can be easily uploaded and watched by others on the web, the community potentially expanded from the classroom to include the entire world. Awareness of the changed nature of the community affected the choice of topic as when, for example, Corine wanted to show how feminism had impacted her life. Cora, aware of her expanded potential audience, made a clear choice to use "less formal language and more emotional words [than in our previous essays]." Theresa was aware that for an oral narrative to be posted online, she had to create an engaging narrative "*que captara la audiencia, la cual incluiría más que mis compañeros de clase*" 'that captured the audience, who would include more people than those of my classmates.' In addition to presenting the topic to the class, Theresa "carefully searched for the right images and music to influence the audience's feelings. In her statement "*ojalá mis lectores puedan sentir las mismas emociones que sentí yo al cruzar la frontera al lado mexicano*" 'I wish my audience could feel the same emotions that I felt when I crossed the border to the Mexican side,' she made it clear that the presence of an audience was influencing her digital and linguistic choices.

Finally, the division of labor included both vertical and horizontal relationships. Despite the DS being an individual assignment, the participants worked collaboratively to develop the foundation of the original essays through the discussion board and class talk; in some instances, learners also worked collaboratively to write the essays. After writing the narrative for the DS separately, learners worked together in the "circles," which became a forum to review scripts and to suggest changes to them. For example, Cora mentioned in her final reflection piece that

> *los círculos" en el proceso de creación me han ayudado en el proceso de reescribir el guión [...] Una ventaja de escuchar a mis compañeras sobre las historias digitales era escuchar los temas de sus historias.*"

'Using the circles in the creation stage helped me in the process of rewriting my script [...] An advantage of listening to my classmates about their stories was to listen the topics of their stories.'

Involvement in the circles was clearly helpful not only for the writing process but also in relation to the selection of images and sound. As Corine commented in her final reflection, *se trata de un método ventajoso porque el conocimiento de que mis compañeros van a revisar mi trabajo en detalle, me dio ganas de trabajar aún más para asegurarse de que todo era perfecto.*

'this method works well because knowing that my classmates were going to revise my work in detail gave me the motivation to work even harder to ensure that everything was perfect.'

While still waiting for the instructor's feedback (the vertical relation), the knowledge that they would be providing and receiving ongoing feedback in their groups encouraged learners to develop interesting narratives.

4.2 Research Question Two

The second research question focused on the linguistic and structural writing similarities and differences between traditional forms of literacy and 21st-century literacies (digital stories), the six participants were positive that their linguistic ability had improved during the creation of the DS. They made reference to acquiring a wider range of vocabulary, being able to use connectors better and work with different registers (formal, expressed in longer sentences for the essays, and less formal, shorter sentences for the DS). For example, Susan also explains that the DS was "less formal, shorter sentences, but still wide range of vocabulary, and mixing up of connectors. I think the same amount of planning went into it, but definitely less research!" The DS genre allowed the participants to explore other vocabulary not studied in class due to the oral nature of the task and permitted the integration of sound and image to make the narrative more dynamic.

Regarding the structure of writing, some of the participants approached the writing of the DS as they had done in previous assignments. Cora mentioned that "generally the writing process was the same in terms of revision for aspects such as repetition, grammar, etc." and Soraya added, "it was very similar to the academic style in that we still needed to have basically the same structure with the definition and the thesis." The script of the digital story, however, was "more of a narrative and so it was structured differently in that the objective of the writing had to be specified more in the form of a moral of a story rather than the classical thesis statement."

With respect to grammar, Theresa felt that the structure of the DS "offered much more creative liberty, because the thesis did not have to be explicitly stated in the introduction [...] but could be expounded upon throughout the whole video. Also, the tone of my digital story was much more casual than any academic essay." Corine noted that it felt strange to be incorporating personal content: "Since I'm so used to writing technical pieces, it was kind of hard to incorporate something personal into the assignment. The academic style was relatively similar—with transitional phrases and the like."

Other comments about structural differences were mentioned in the final reflections. The participants recognized that narrative writing was different. For example, Cora mentioned in her final reflections that

> *En contraste a los ensayos tradicionales como la exposición y la argumentación, la narración final es muy diferente. Ella tiene un aspecto personal mientras los ensayos solo desean hechos. Los sentimientos del escritor no valen nada en un ensayo tradicional.*

'In contrast with traditional essays like the exposition and argumentation, the final narration in very different. It has a personal aspect, while the essays only want facts. The writer's feelings are irrelevant in a traditional essay.'

Corine also noted the difference between academic writing and the DS when she explained,

> *Cuando empecé a redactar, descubrí que el proceso de escribir no era similar a lo que estaba acostumbrada. Por ejemplo, a lo largo del curso, hemos escrito escrituras en los estilos tradicionales, como la exposición y la argumentación, sino por la historia digital, el guión se trataba de un evento que me impactó personalmente, como una narrativa, mientras que los otros estilos son más objetivas y centrados en la investigación. Además, el guión es el único que se escucha como audio en vez de leerlo como un ensayo tradicional, por lo que necesitaba que tener en cuenta aspectos tales como el tono y el ritmo de mi voz.*

'When I started composing, I discovered that the process of writing was not similar to what I was accustomed to. For example, during the course we have written essays in the traditional styles, such as exposition and argumentation, but for the DS, the script was about an event that impacted me personally, like a narrative, while the other styles were more objective and centered in an investigation. Besides, the script is the only one that can be heard as audio instead of reading it is like a traditional essay, that is why I needed to take into account aspects such as the tone and the rhythm of my voice.'

In her reflection, Matilde realized the importance of the oral nature of the DS and how the use of sound and image changed the way she approached the writing:

> *Tradicionalmente, ofrecíamos los mensajes en forma escrita bajo las modalidades de exposición o argumentación, y quienes tenían la habilidad de hacer magia con las palabras, podían graficar con las letras sus motivaciones o emociones. Con la historia digital, aquella magia ya no es una limite. Usando de creatividad tendremos la posibilidad de visualizar, dar dinamismo y audio a nuestras ideas haciendo de ellas una obra de arte.*

'Traditionally, we offered messages in written form under the modalities of exposition and argumentation, and those who had the ability to make magic with words could express with written words their motivations and emotions. With the DS, that magic is not a limitation. By using creativity, we have the possibility to visualize, give dynamism and audio to our ideas transforming them into a work of art.'

4.3 Research Question Three

Question 3 examined learners' perceived benefits or drawbacks when developing DS assignments. When the participants were asked in the questionnaires whether the DS promoted their writing development, five of the participants agreed. Cora, however, did not enjoy this activity as much because "for someone who doesn't want to put her personal life into her school assignments, the digital stories do nothing for me; especially since the sound of my own voice (and other people's voices) makes my skin crawl." The other five participants, however, agreed that the DS assignment did develop their writing skills. Matilde, for example, thought it a "great instrument to improve writing skills. We have to write in different styles, to cover a topic within certain minutes." In terms of linguistic components, Soraya mentioned that "[DS] really makes you focus on the tone and rhythm of your writing and the descriptions to evoke emotion and imagery. Vocabulary in this case is key." Susan also commented that working on the DS helped develop narrative writing skills in a "fun project [that] challenged students to make their stories interesting, educational, and purposeful for a real audience." Overall, learners felt that the new strategy helped them foster their language development. Soraya, for example: "[The DS is] a different type of writing technique and requires a different thinking process." An additional benefit was in improving technical literacy. Theresa mentioned that "It improves technology literacy, and it's a refreshing change from the PowerPoint presentation!" Cora, however, commented that "*había desafíos de aprender, comprender, y adaptarme a esta nueva forma de expresión escrita/oral. Fueron la incorporación del aspecto personal al guión, la grabación de la narración, y el trabajar con un software desconocido anteriormente.*" 'there were challenges when I was learning, comprehending, and adapting to this new form of oral/written expression. These [challenges] were the incorporation of the personal aspect of the script, the recording of the narration, and working with an unknown software.'

However, overall, despite her initial problems, Theresa later posted,

> *Al final, me divertí mucho trabajando en este proyecto. Estoy orgullosa de poder manejar un nuevo programa y haber creado un video que combina una narración con imágenes y música complementarias. Aunque hubo momentos en los cuales pensaba que no iba a poder hacerlo, lo hice. En general, la historia digital me ha servido como una buena culminación de todo el trabajo de Español 401, y me ha hecho más instruida sobre la tecnología educativa del siglo XXI.*

'By the end, I was enjoying working with this project very much. I am proud of being able to manage a new program and to have created a video that combines narration with images and complementary music. Although there were moments in which I thought that I would not be able to do it, I did. In general, the DS has served me as a good culmination of all the work in Spanish 401, and has helped me to be better informed about the educational technologies of the 21st century.'

Corine felt positive, too.

Como estudiante que aprende las lenguas extranjeras, he tenido que producir varias escrituras en los últimos años; sin embargo, ninguna de ellas ha sentido tan poderosa o tenía la calidad de vida como esta obra logró. Con imágenes vibrantes, música intensa, y diálogo impresionante, las piezas de la historia digital combinan para formar una visión inspiradora del feminismo, en la cual la retroalimentación y las revisiones han servido para refinarla a la excelencia. Por lo tanto, a pesar de que fue lo más desafiante, este proyecto fue lo más satisfactorio porque era más que una escritura; era una ventana a mi pasado, a mi historia.

'Like any students learning a foreign language, I have had to produce several writing assignments in the last few years; however, none of them have felt as powerful or had such an impact on my life as this one did. With vibrant imagery, intense music, and an impressive dialogue, the pieces of the DS combined to form an inspired vision of feminism and all the feedback and revisions have helped to refine it to excellence. Despite its challenges, this project was very rewarding because it was more than writing; it was a window to my past, to my history.'

Susan concluded,

No soy una person creativa de manera digital, entonces estoy feliz que está hecha ya mi historia y que esta experiencia está terminada. Sé que otras personas se divirtieron, pero no me diverti en trabajar con el programa. Siempre resulta más difícil para mi usar la tecnología.... Sin embargo, me alegro de tener esta experiencia de hacer la historia digital y de dar crédito a mis queridos amigos que salen en la historia.

'I am not a digital creative person, so I am happy that my story and that this experience are over. I know that other people had fun, but I did not enjoy working with the program. It is always difficult for me to work with technology. However, I am happy to have had the experience of creating a DS and give credit to my dear friends that appear in my story.'

5. Discussion

Given the growing interest in helping learners understand digital literacies and become proficient writers in the 21st century, the purpose of this study was to explore learners' perceptions about the integration of digital literacies in an advanced Spanish writing course. Data collected through questionnaires, journals, and reflections provided evidence of learners' goal-oriented actions as they developed content and negotiated the structure of the DS using in-class discussions, their previous essays, collected images and sounds, and the script. The data also highlighted how learners compared the DS with previous more traditional writing assignments and how they judged the benefits of including the DS in their advanced FL writing curriculum.

The first research question concerned learners' interactions within an activity system on computer-mediated Spanish writing environment. The results of the study confirm that the type of tools and artifacts employed in a FL writing task are not neutral, and thus, the integration of a tool such as Final Cut shaped the writing process in a distinctive manner (Werstch, 1991). The digital story was mediated by specific tools (Final Cut, images, sounds, previous essays), which encouraged learners to orient and reorient their actions in relation to the object and the desired outcome to become effective communicators in the 21st century and also pushed them to learn a new set of multimodal communicative conventions. Echoing Engeström (1999), the use of these tools had a direct effect on the interrelationship among the components of an activity system. By including the possibility of uploading the digital story to the web, the community expanded to include not only the class participants, but also anyone who had access to YouTube. Becoming aware of this expanded audience influenced the choice of language, narrative, images, and sound learners could use in order to evoke appropriate emotion in their potential audience. In their interactions with the text, learners interpreted and negotiated a new set of rules, requiring integration of narrative, images, and sounds. Learners also welcomed new operational rules such as the introduction of "circles" to encourage the giving and receiving of feedback as they created their scripts and worked to integrate images and sound within the narration. While still looking forward to the instructor's comments, the majority of learners enjoyed the shared responsibility of providing comments to one another.

Learners used, created, and modified a broad range of "material-semiotic artifacts," that linked longer-term processes to shorter-terms events (Lemke, 2000). While the learners completed the DS at the end of the semester, the task-based approach in the classroom allowed learners not only to focus on goal-oriented activities such as selecting images, integrating sound, or thinking about the topic, but also to have time to reflect on the different phases of the DS development. Although there was a shared goal of completing the DSs, the data analyses suggest, as Leontiev (1978) pointed out, that the learners performed an "action" in relation to their goal under differing conditions (e.g., analyzing images in the discussion boards, script development, etc.), and the way in which they carried it out varied depending on how they interpreted the writing task. The instructor had provided the tools, tailored to best assist them to achieve the goal, but learners' interpretations of the object (the creation of their DS) influenced the use they made of these tools. The tools also mediated learners' evolving perception of the object, and their use helped learners to focus on different aspects of the DS in a sophisticated way: hence, they operationalized the activity to suit their own understanding.

In regard to the second question, how learners perceived similarities and differences between traditional forms of literacy (i.e., genres such as argumentative and expository essays) and 21st-century literacies (i.e., digital stories), the results presented here support previous research that highlights the relevance of the written component of the DS. As described by Gakhar and Thompson (2007) and Robin (2006), while searching for images and sounds, the learners maintained focus and constructed narratives, organizing their thoughts in coherent storylines in a project "that was more than a writing task" (Corine, reflection). Learners often struggle to understand the differences between the argumentative and expository essays. Because it was student's initial encounter with a DS, it took several drafts had been written and discussed in the "circles" for some learners to become comfortable with the new genre. Ultimate success was perhaps due to the task-based approach, which allowed the participants enough time to reflect on difficulties and prepare for the final DS product. After working on several assignments that required them to provide facts devoid of emotions, the personalized nature of DSs also advanced their understanding of the differences between writing genres. In terms of language, despite the more "casual" quality of DSs, (Reyes Torres et al., 2012), learners followed the rules of standard written language, showing a similar concern for grammar that they had learned and applied to their previous academic essays. As suggested by Gregori-Signes (2008), while focusing on the written language, learners became aware of the oral story-telling nature of the script and learned to include appropriate stylistic devices to suit the narration. Thus, by developing the DSs, learners learned to select and work with different tools for diverse communicative acts (writing and oral aspects of the DS), which have become part of the 21st-century communicator.

Finally, the last question looked at the learners' perceived benefits or drawbacks of the DS as a writing development tool. Overall, except for one student who was not keen to share her personal experiences with a wide audience, participants were positive about the inclusion of the DSs in the FL writing class in terms of linguistic and genre-related development and technological literacy. The perceived value of the DS largely came from having a wider audience for their work; writing was no longer an act between the instructor and the learner, but had become a communicative act between the learner and a wider audience. To gain that audience's attention, learners created interesting, educational, and purposeful stories. To do so, they needed grammar to convey the desired tone and rhythm, and also the right vocabulary to evoke emotions.

Interestingly, several learners commented on their use of Final Cut during the final stages of the DS development. Given that this was their first time working with the tool, its introduction was definitely an additional challenge. Writing the story, searching for the images, selecting the music, and combining all the components in the software were ambitious processes. Yet, after overcoming the initial problems, learners welcomed the change from traditional classroom tools and learned to value these 21st-century tools as they became more adept users.

6. Conclusion

For our learners to become 21st-century writers, they need to use 21st-century tools in an effective and pedagogically sound manner. As seen in this study, the tools are powerful in the mediation of knowledge. The introduction of Final Cut to manipulate images, sounds, language, and artifacts influenced how learners viewed and understood the writing task, and consequently how they viewed the audience to which their work was directed. The presence of a wider audience encouraged

learners to search for discourse, vocabulary, and a style that could best transmit their message, promoting a strong sense of authorship.

The introduction of the DS into our course allowed learners to extend their knowledge of genres, grammar, and vocabulary; they also completed learning tasks that could foster personal creativity and practice with specific discourse structures. All of these factors earn the DS a place in our FL curriculum, especially since it can allow learners to experiment with other writing genres in the future. The personalized nature of the DS led some participants to view their work differently. Because their stories came from a personal standpoint, learners pondered topics that could enrich the dialogue between writers and a new much wider audience. As Daley (2003) stated, "like text, multimedia can enable us to develop concepts and abstractions, comparison and metaphors, while at the same time engaging our emotional and aesthetics sensibilities" (p. 34).

Despite the study's positive results, we must be cautious because this preliminary study included only six participants. This small sample size affects generalizability of results. Yet, a small study like this one can provide a "significantly enhanced awareness of the contextual and interactional dimensions of language use, and increased 'emic' (i.e., participant-relevant) sensitivity toward fundamental concepts" (Firth & Wagner, 2007, p. 757) and can acknowledge the importance of social factors in the acquisition process (Lafford, 2007). Another limitation of the study is that it focuses on the learners' perspectives rather than their writing performance. Given the close connection between the traditional narrative essay and the DS narrative, research examining these two genres could provide insightful information regarding the two modes (traditional and 21st-century-DS format) of the narrative essay. Finally, from a pedagogical perspective, we believe that the DS technique could reshape not only the types of tasks set in FL advanced writing courses, but also open the writing class to 21st literacy practices.

References

Alcantud Díaz, M. (2008). Experiencia en filología inglesa: Portafolio electrónico y relato digital. Valencia. Open ACS. *Jornadas sobre el uso de software libre en educación superior*. Server dínformatica, vicerectorat de postgrau y vicerectorac de convergencia Europes i qualitat de la Universitat de València. Retrieved from http://www.uv.es/anglotic/maldiaz/documents/open_ACS_08_final__paper.pdf

Alcantud Díaz, M. (Ed.). (2010). *Tales in two minutes: Cuentos en dos minutes: NTIC and Project Work*. Valencia: Reproexpres ediciones. Retrieved from http://www.uv.es/anglotic/maldiaz/documents/Tales_final.pdf

Blin, F., & Appel, C. (2011). Computer supported collaborative writing in practice: An activity theoretical study. *CALICO Journal, 28*(2), 473–497.

Castañeda, M. (2013). "I am proud that I did in and it's a piece of me": Digital storytelling in the foreign language classroom. *CALICO Journal, 30*(1), 44–62.

Castañeda, M., & Rojas-Miesse, N. (2013, May 24). *Using digital stories to enhance and develop L2 language skills.* Paper presented at CALICO, Manoa, Honolulu: HI.

Daley, E. (2003). Expanding the concept of literacy. *EDUCAUSE Review, 38*(2), 32–40

Ellis, R. (2003). *Task-based language learning and teaching.* Oxford: Oxford University Press.

Elola, I., & Oskoz, A. (2011). Writing between the lines. Acquiring the presentational mode through social tools. In N. Arnold, & L. Ducate, (Eds.), *Present and future promises of CALL: From theory and research to new directions in language teaching,* (pp. 171–210). San Marcos, TX. CALICO.

Engeström, Y. (1999). Innovative learning in work teams: Analyzing cycles of knowledge creation in practice. In Y. Engeström, R. Miettinen, & R. Punamäki (Eds.), *Perspectives on activity theory* (pp. 377–404). Cambridge: Cambridge University Press.

Firth, A., & Wagner, J. (2007). On discourse, communication, and (some) fundamental concepts in SLA research. *Modern Language Journal, 91,* 757–772.

Flower, L., & Hayes, J. R. (1981). A cognitive process theory of writing. *College Composition and Communication, 32*(4), 365–387.

Gakhar, S., & Thompson, A. (2007). Digital storytelling: Engaging, communicating, and collaborating. In C. Crawford et al. (Eds.), *Proceedings of the Society for Information Technology & Teacher Education International Conference 2006* (pp. 643–650). Chesapeake, VA: AACE.

Gregori-Signes, C. (2008). Integrating the old and the new: Digital storytelling in the EFL language classroom. *GRETA, 16,* 43–49.

Haffner, C. A., & Miller, L. (2011). Fostering learning autonomy in English for science: A collaborative digital video project in a technological learning environment. *Language Learning & Technology, 15*(3), 68–86.

Hampel, R., & Hauck, M. (2006). Computer-mediated language learning: Making meaning in multimodal virtual learning spaces. *The JALT CALL Journal, 2*(2), 3–18.

Kajder, S. B., & Swenson, J. A. (2004). Digital images in the language arts classroom. *Learning and Leading with Technology, 31*(8), 18–21.

Kessler, G. (2009). Student-initiated attention to form in wiki-based collaborative writing. *Language Learning & Technology, 13*(1), 79–95.

Lafford, B. (2007). Second language acquisition reconceptualized? The impact of Firth and Wagner (1997). *Modern Language Journal, 91,* 735–756.

Lambert, J. (2002). *Digital storytelling: Capturing lives, creating community.* Berkeley, CA: Digital Diner.

Lantolf, J. P., & Thorne, S. L. (2006). *Sociocultural theory and the genesis of L2 development*. Oxford: Oxford University Press.

Lemke, J. L. (2000). Across the scales of time: Artifacts, activities, and meanings in ecosocial systems. *Mind, Culture, and Activity, 7*(4), 273–290.

Leontiev, A. (1978). *Activity, consciousness and personality*. Englewood Cliffs, NJ: Prentice Hall.

Merriam, S. B. (1998). *Qualitative research and case applications in education: Revised and expanded from case study research in education* (2nd ed.). San Francisco, CA: Jossey-Bass.

Ohler, J. (2006). The work of digital storytelling. *Educational Leadership, 63*(4), 44–47.

Ohler, J. (2008). *Digital storytelling in the classroom: New media pathways to literacy, learning, and creativity*. Thousand Oaks, CA: Corwin Press.

Oskoz A., & Elola, I. (2012). Understanding the impact of social tools in the FL writing classroom: Activity theory at work. In G. Kessler, A. Oskoz, & I. Elola (Eds.), *Technology across writing contexts and tasks* (pp. 131–154). San Marcos, TX. CALICO.

Prabhu, N. S. (1987). *Second language pedagogy*. Oxford, UK: Oxford University Press.

Rance-Roney, J. (2008, March). Digital storytelling for language and culture learning. *Essential teacher*. Retrieved from http://www.nwp.org/cs/public/download/nwp_file/12189/Judith_Rance-Roney_Digital_Storytelling.pdf?x-r=pcfile_d

Reid, M., Parker, D., & Burn, A. (2002). Digital video report, BECTA. Retrieved from partners.becta.org.uk/upload-dir/downloads/page_documents/research/ dvreport_241002.pdf

Reinders, H. (2011). Digital storytelling in the foreign language classroom. *ELT-WO Journal, 3*, 1–9.

Reyes Torres, A., Pich Ponce, E. & García Pastor, M. D. (2012). Digital storytelling as a pedagogical tool within a didactic sequence in foreign language teaching. *Digital Education Review, 22*, 1–18.

Richards, J., Platt, J., & Weber, H. (1985). *Longman dictionary of applied linguistics*. Harlow, UK: Longman.

Robin, B. (2006). The educational uses of digital storytelling. In D. A. Willis, J. Price, N. E. Davis, & R. Weber (Eds.), *Proceedings of the Society for Information Technology & Teacher Education International Conference 2006* (pp. 709–716). Chesapeake, VA: AACE.

Robin, B. R. (2007). The convergence of digital storytelling and popular culture in graduate education. In R. Carslen et al. (Eds.), *Proceedings of Society of Information Technology and Teacher Education International Conference 2007* (pp. 643–650). Chesapeake, VA: AACE

Robin, B. (2008). The effective uses of digital storytelling as a teaching and learning tool. In J. Flood, S. Health, & D. Lapp. (Eds.), *Handbook of research in teaching literacy through the communicative and visual arts* (pp. 429–440). New York, NY: Lawrence Erlbaum Associates.

Sadik, A. (2008). Digital storytelling: A meaningful technology-integrated approach for engaged student learning. *Education Tech Research and Development, 56*. Retrieved from http://classroomweb20.pbworks.com/f/digital+storytelling.pdf

Samuda, V., & Bygate, M. (2008). *Tasks in second language learning*. Basingstoke, England: Palgrave Macmillan.

Skehan, P. (1998). *A cognitive approach to language learning*. Oxford, UK: Oxford University Press.

Thorne, S. L. (2003). Artifacts and cultures-of-use in intercultural communication. *Language Learning & Technology, 7*(2), 38–67.

Vinogradova, P. (2008). Digital stories in an ESL classroom: Giving voice to cultural identity. *LLC Review, 8*(2), 56–71.

Vinogradova, P. (2011). *Digital storytelling in ESL instruction: Identity negotiation through a pedagogy of multiliteracies*. (Unpublished doctoral dissertation). University of Maryland Baltimore County, Baltimore, MD.

Vinogradova, P. (2014). Digital stories in heritage language education: Empowering heritage language learners through a pedagogy of multiliteracies. In T. Wiley, D. Christian, J. K. Peyton, S. Moore, & N. Liu (Eds.), *Handbook of heritage, community, and Native American languages in the United States: Research, educational practice, and policy* (pp. 314–323). New York: Routledge.

Vinogradova, P., Linville, H. L., & Bickel, B. (2011). "Listen to my story and you will know me": Digital stories as student-centered collaborative projects. *TESOL Journal, 2*(2), 173–202.

Vygotsky, L. (1978). *Mind in society*. Cambridge, MA: Harvard University Press

Werstch, J. V. (1991). *Voices of the mind*. Cambridge, MA: Harvard University Press.

Willis, J. (1996). *A framework for task-based learning*. London, England: Longman.

Chapter 9

Exploring the Affordances of Digital Social Reading for L2 Literacy: The Case of eComma

CARL S. BLYTH
University of Texas at Austin (USA)

Abstract

Current e-reading devices allow readers to annotate a text and to share those annotations with others. The result is a new literacy practice—digital social reading—that violates many teachers' expectations of what it means to read based on their shared "print culture" (Baron, 2013). This study focuses on teachers' perceived pedagogical affordances of a web-based application for digital social reading. An analysis of the use of a web application in four L2 classrooms suggests how teachers may employ digital social reading as a "bridging activity" (Hayles, 2012; Thorne & Reinhardt, 2008) to resolve the apparent clash between print and digital cultures. This essay argues that digital social reading is best understood in terms of a "participatory culture" (Jenkins, 2009) in which digital practices are transforming individual practices into social ones.

1. Introduction

Many teachers conceptualize reading as a private act that may or may not be publicly shared. In such a view, the reader is an individual who interprets a text by herself and decides whether to share her interpretation in a public forum such as a classroom discussion. However, online reading is changing our understanding of what it means to read by blurring the line between private interpretation and public discussion. The history of marginalia provides a good illustration. During the 18th century, people routinely marked up books or novels to give as personalized gifts (Jackson, 2001). This practice was intended to give the reader the sense of socializing with another person during the private act of reading.

Sam Anderson (2011), book critic for *The New York Times*, has called today's use of digital marginalia "a natural bridge" between the reader's private and public worlds (p. 1). Likening the practice of digital marginalia to Twitter, Anderson envisions the day when readers will instantaneously share their thoughts with others about whatever they are reading.

> I've long been frustrated with the "distance" between criticism and reading itself. Most critical energy is expended in big-picture work—situating texts in history, talking about broad themes—all of which is useful but hardly touches the excitement of actual reading, a process of discovery that happens in time, moment by moment, line by line. What I really want is someone rolling around in the text. I want noticing. I want, in short, marginalia, everywhere, all the time. (Anderson, 2011, p. 1)

It appears that Anderson's wish to be "rolling around in the text" (2011, p. 1) with others has already come true. Thanks to the rapid rise of e-readers and tablets, digital marginalia is returning to its social origins. Today, most e-reading devices allow users to annotate a text and to share those annotations with incredible speed and reach. The result is a new literacy practice called *digital social reading*. To ensure that this new literacy practice remain easy, personal, and open, activist James Bridle created the Open Bookmarks website in 2011. On the home page of his website, Bridle argues that for social reading to belong "to the reader and nobody else," an individual's annotations must be "shareable, saveable and persistent." Bridle illustrates digital social reading with multiple examples, including one that indicates how teachers are exploring its pedagogical affordances: "A teacher makes a number of annotations in an ebook. She exports them, sends them to her students, who import them into their own copies of the book" (Bridle, n.d., What is Social Reading?, Some examples of Social Reading, No. 3).

Despite the growing interest in classroom applications of digital social reading, some educators remain skeptical. For these teachers, the use of collaborative learning methods to teach a private, mental skill seems paradoxical if not downright misguided (Bauerlein, 2008, 2011). These skeptics typically frame social reading in terms of deficiency, that is, as deficient readers relying on the literacy skills of others. Therefore, it appears that the perceived affordances of digital social reading—as with any new technology or practice—depend largely on one's understanding of what it means to read. The relational term *affordance*, a central feature of ecological approaches to learning and human development, is defined as "relations of possibility between animals and their environment" (Neisser, 1987, p. 21). Following Gibson (1977, 1979), van Lier (2004) describes the concept for the L2 learning environment as "relationships that provide a 'match' between something in the environment ... and the learner" (p. 96) or "what is available to the person to do something with" (p. 91). The main point is that pedagogical affordances are not inherent in objects but must be perceived by an agent—either a teacher or a learner. In brief, this essay examines how L2 teachers perceive the affordances of digital social reading with which they "do something" to improve their praxis. This essay focuses on four case studies of the use of a web-based application for digital social reading called eComma. Taken together, these case studies illustrate digital social reading as a set of emergent literacy practices that are redefining what it means to read in the 21st century.

This chapter is divided into five sections. In the first section, digital social reading is shown to violate teachers' cultural expectations of what it means to read based on their shared "print culture" (Baron, 2013). It is argued that any perceived affordances of digital social reading are directly related to how teachers understand their own literacy practices. In the second section, various forms of social reading are defined, and social reading tools, including eComma, are described and compared. In the third section, studies on collaborative forms of L1 reading and textual annotation are reviewed. Still in its infancy, research on digital social reading in the L2 context is briefly summarized. In the fourth section, first-person accounts of the classroom use of eComma are analyzed to capture selected L2 teachers' perceptions of the affordances of this new literacy tool. An analysis of four case studies suggests ways that digital social reading may be used as a "bridging activity" (Hayles, 2012; Thorne & Reinhardt, 2008) to help resolve the apparent clash between print and digital cultures. The chapter concludes with suggestions for future research on digital social reading in light of *participatory culture*, a new framework for understanding the role of new media and collaborative social action (Jenkins, 2006, 2009).

2. Reading as Cultural Practice

Many forms of digital literacy—including digital social reading—remain highly controversial, figuring prominently in the debates concerning American students' reading skills (Bauerlein, 2008, 2011; Carr, 2010). In a nutshell, critics of digital literacy view today's online readers as distracted and hyperactive, unable to focus long enough to unravel the complexities of a literary text. Critics worry that reliance on crowd-sourced commentary—a prominent feature of social reading—will contribute to the further deterioration of traditional literacy prized by humanists who associate close reading skills with print (Hayles, 2012).

According to media scholar Katherine Hayles (2012), such criticism fails to acknowledge that digital literacy practices have real cognitive benefits, such as enabling readers to sort more information more quickly. More important, Hayles notes that such criticism ignores the crucial fact that digital media are indispensible tools for "how we think" in the Digital Age: "The Age of Print is passing, and the assumptions, presuppositions, and practices associated with it are now becoming visible as media-specific practices rather than the largely invisible status quo" (2012, p. 2). Hayles argues that the shift from print to digital is having far-reaching effects on higher education, especially on the humanities: "Starting from mind-sets formed by print, nurtured by print, and enabled and constrained by print, humanities scholars are confronting the differences that digital media make in every aspect of humanistic inquiry, including, conceptualizing projects, implementing research programs, designing curricula, and educating students" (2012, p. 2).

Hayles's comments emphasize the importance of understanding reading as a set of historically situated practices and beliefs. Along similar lines, Elizabethan scholar and cultural historian Thomas Pettitt (2007) coined the term "Gutenberg Parenthesis" to refer to the period in Western history within which various forms of writing—essays, letters, plays, poems—were assumed to be the original products of a single author. Pettitt uses the well known grammatical device to capture the historical development of literacy in terms of an interruption, thereby highlighting the continuity between what comes before with what comes after. According to Pettitt, during the Gutenberg Parenthesis, literacy practices came to be increasingly understood in terms of individual readers and writers. In contrast, "pre-parenthetical" and "post-parenthetical" literacy practices emphasize collaborative forms of textual composition and interpretation. The practices associated with Pettitt's pre- and postparenthetical periods are similar to Walter Ong's concepts of primary orality and secondary orality. In brief, Ong (1982) draws parallels between the primary oral practices in preliterate cultures with today's hybrid forms of computer-mediated communication, "essentially a more deliberate and self-conscious orality, based on the use of writing and print" (p. 135).

Baron (2013) notes that the development of print culture took several centuries and entailed shifts in beliefs and attitudes about what it meant to read. Historically, the act of reading was either performed aloud for listeners or performed *sotto voce* for oneself. As such, carrels became a common architectural feature in 12th-century cloisters to keep mumbling monks from disturbing each other. Baron points out that it was not until print culture started to coalesce in the late 14th-century that silent, private reading became a common practice: "although reading aloud to others—both literate and nonliterate audiences—was common historically, reading strictly by oneself became the norm by the time print culture was established" p. 205).

Baron (2013) claims that the durability of the printed text was a defining feature in print culture's understanding of what it meant to read. For example, the printing press turned language—a heretofore oral and, therefore, ephemeral phenomenon—into a durable text. The durability of the printed text allowed it to be passed around and shared, which in turn promoted the rise of other common literacy practices such as marginalia. For example, books passed down within a family routinely contained annotations made by members who had previously read the text. The assumption behind such shared annotation behavior is the possibility "for someone to subsequently return to that text" (p. 205).

Baron (2013) argues that digital literacy, or, as she puts it "reading onscreen" (p. 206) presents considerable challenges to teachers' understanding of reading. In particular, Baron cites the common digital practice of multitasking as "radically altering our understanding of what it means to attend to a text" (p. 208). She exemplifies multitasking in these terms: "searching for inexpensive plane fares while reading, changing one's Facebook status while reading, sending and receiving text messages while reading ..." (p. 208). From these examples, it is apparent that Baron equates multitasking with an interruption in a reader's attention, the moment when a reader turns away from the text and is therefore "off task." And yet, in the Digital Age, a student who turns away from the text to chat with a friend may simply be looking for relevant information to understand the text, the digital equivalent of consulting a dictionary.

According to Hayles (2012), the real problem is that teachers equate reading with a narrowly defined and historically situated literacy practice—the close reading of a *printed text*. Hayles contends that teachers must understand that whenever digital readers leave the actual text to access relevant information elsewhere on the internet, they are still reading. The key, she claims, is to consider different types of computer-mediated semiotic activity as different forms of reading—close reading, hyper reading, and machine reading. Hayles associates each type of reading with a different cognitive mode of information processing:

> Close reading correlates with deep attention, the cognitive mode traditionally associated with the humanities that prefers a single information stream, focuses on a single cultural object for a relatively long time, and has a high tolerance for boredom. (Hayles, 2012, p. 12)

> Hyper reading is a strategic response to an information-intensive environment, aiming to conserve attention by quickly identifying relevant information, so that only relatively few portions of a given text are actually read. Hyper reading correlates with hyper attention, a cognitive mode that has a low threshold for boredom, alternates flexibly between different information streams, and prefers a high level of stimulation. (Hayles, 2012, p. 12)

> Although machine reading may be used with a single text and reveal interesting patterns, its more customary use is in analyzing corpora too vast to be read by a single person. (Hayles, 2012, p. 74)

From Hayles's perspective, such correlations suggest a need for a new reading pedagogy based on the affordances of these different cognitive modes. In fact, she explicitly calls for building bridges between the different types of digital reading so that students can learn how to integrate them during their own sense-making activities (2012, pp. 11–12). Within this new paradigm of literacy pedagogy, "good teachers deliberately focus on what the reader can do, make sure that both teacher and student recognize and acknowledge it, and use it as a platform of success from which to build" (Laurence, 2008, as cited in Hayles, 2012, p. 4). Similarly, Thorne and Reinhardt (2008) propose "bridging activities" for the L2 classroom that are meant to raise learner awareness of the grammatical and lexical structures of their digital vernacular genres such as instant messaging, blogs, wikis, and gaming discourse. Bridging activities such as those proposed by Hayles and Thorne and Reinhardt take students' own digital literacy practices as a natural point of departure for literacy instruction. Such an approach is consonant with the digital social reading tool eComma that was conceived as a bridge between what the reader can do (i.e., digital literacy) and what the reader can not yet do (i.e., close reading of the printed literary text).

3. Defining Digital Social Reading

As noted above, the practices of digital social reading may violate many teachers' understanding of what it means to read. But what exactly are the practices that define this new form of literacy? In brief, digital social reading is the act of sharing one's thoughts about a text with the help of tools such as social media networks and collaborative annotation. The prototypical example is a synchronous activity based on a single, shared text. Luks (2013) defines this type of reading as an "Internet-based activity in which a group of people collaboratively reads, annotates and comments upon a shared text ...; in more language-teaching parlance, one could say that it constitutes a during-reading activity" (p. 8). Luks's definition is akin to Anderson's "process of discovery" that happens "line by line" (2011, p. 2). However, social reading also exists in various asynchronous formats, exemplified by websites such as Goodreads (http://www.goodreads.com/). Such websites serve as meeting places for readers to form online book clubs, essentially digital social networks based on shared literary interests. However, social reading is not limited to people reading the same text. For example, the website Ponder (http://www.ponder.com/) allows students to read different online texts about similar themes and to share their comments with each other in an organized fashion. In essence, this form of social reading aggregates comments about a single topic from different readers who are reading different texts.

The following section reviews these different types of social reading according to whether the activity is synchronous or asynchronous and whether the text is shared or not. In addition, the different tools associated with these different forms of social reading are briefly summarized.

3.1 Synchronous Reading of the Same Text

In 2009, a tool called eComma was created to help learners overcome the problems associated with the close reading of literary texts. eComma turns a print text into a digital text so that students can leverage their skills in both hyper reading and machine reading for the purpose of close reading. Designed by faculty and graduate students from the Department of English at the University of Texas at Austin with funds from the National Endowment for the Humanities, the tool was originally developed to enable English literature students to build a collective commentary on a literary text and to search, display, and share the commentary online in a more pliable form. Originally conceived to facilitate the close reading of literature by American college students, the tool was adapted to the second language environment by the Center for Open Educational Resources and Language Learning (http://www.coerll.utexas.edu/coerll/). Currently available as an openly licensed freeware module that plugs into Drupal 7 (an open source content management system), eComma offers a new kind of reading experience: a group of readers can share their reactions to a text instantaneously and build a body of commentary about the text together (Blyth, 2013).

When students log on to the eComma website, they must first choose their course/section and then access a text that appears on the left side of the screen (see Figure 1).

Icons are arrayed in the middle of the screen, between the text and the comments. These icons represent various features of the software: word clouds, tags, comments, and comment clouds. eComma automatically displays all the words in a given text in a word cloud, an iconic representation of frequency—the more frequent the word in the text, the bigger the word in the cloud. Readers can glean important information by quickly glancing at a text's word cloud and discerning keywords. When a reader highlights any part of the text by dragging the cursor over a selected passage, a pop-up window appears prompting the reader to leave a tag or a comment. A tag is a keyword or label. In contrast, a comment is longer than a tag, usually a sentence or two. Comments are usually written text, but may also take multimedia forms such as images or audio files or even videos as shown in Figure 2.

Figure 1
eComma Layout

Figure 2
Multimedia Comments

Comments often prompt replies from other readers during a synchronous reading activity. Thus, a single comment about a highlighted word or phrase may give rise to a threaded discussion. Similar to a word cloud, a comment cloud displays all the words found in all the comments left by a group of readers. Comment clouds are an effective way to ascertain the main ideas of the group's collective commentary.

In addition to these features, eComma contains a list of users and a heat map of annotation activity as shown in Figure 3.

Figure 3
Heat Map

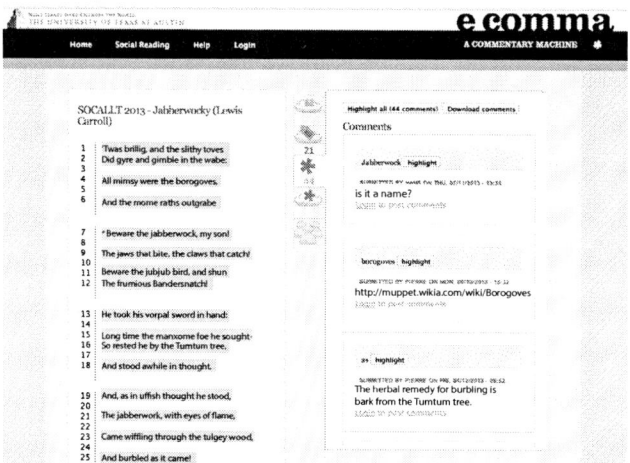

By glancing at the list of users, an instructor can easily discern how many and what kinds of annotations users have made. Heat maps are visual representations of aggregated annotation activity. The more times a word or phrase is annotated, the darker the word or phrase appears in the heat map. Heat maps are handy tools for determining at a glance which parts of the text attract the most attention from the readers. In addition to eComma, there are other similar collaborative annotation tools such as Annotator (Open Knowledge Foundation), eMargin (Birmingham City University), and Classroom Salon (Carnegie Mellon University).

3.2 Asynchronous Reading of the Same Text

A rather different social reading experience is the digital book club that combines the old-fashioned conviviality of the face-to-face book club with the affordances of a digital social network. Two websites that focus on this kind of experience are Goodreads and LibraryThing (https://www.librarything.com/). Both websites bill themselves as places where bibliophiles can meet and do what they like to do best—discuss books. In general, both sites create spaces where users can interact about their favorite books in a variety of asynchronous ways. Users can view each other's libraries of favorite books, read each other's book reviews, and join discussion groups about books they have read or wish to read.

While these websites are undeniably social, they are not social in the same way as eComma, which allows readers to share the synchronous experience of reading. And yet, synchronous and asynchronous reading practices are both routinely described by the same spatial metaphor. In this metaphor, the book or the text is conceptualized as a meeting place where readers can talk and socialize (Howard, 2012). For example, a popular social reading website Readups (http://www.readups.com/), a play on the popular website MeetUp (http://www.meetup.com/) describes the social reading experience as *"meeting up* inside a book." This metaphor highlights the virtual nature of social reading, an activity that is shared yet spatially and cognitively distributed.

3.3 Reading Different Texts as a Group Activity

Yet another form of digital social reading includes groups of people who follow each other's comments while reading different online texts. With the Ponder browser plugin, a teacher can manage open-ended, self-directed reading assignments during which students are allowed to search the internet for relevant texts. A typical Ponder assignment would be to have students search for media examples of "insincere political apologies" or "conflict resolution" or "correlation interpreted as causality." Unlike most free-response annotation tools such as eComma, Ponder requires students to annotate texts with prefabricated *microresponses* that are aggregated and visualized for the group to analyze. This form of social reading allows students to search and browse the internet in an autonomous fashion. As such, this kind of social reading is said to encourage intentional sharing (i.e., ponder, then share) since students must ask themselves whether the texts that they choose to read and comment upon will be interesting and relevant to their classmates.

Ponder affords a form of social reading more akin to metareading, that is, reading comments about reading. Crucial to Ponder's metareading function is the microresponse. According to Ponder's research, social readers exhibit a strong preference for short, direct comments, often refusing to read anything longer than a sentence (A. Selkirk, Ponder CEO, personal communication, November 7, 2013).[1] To encourage students to "read each other," Ponder limits responses to short phrases called *sentiments* that fall into three categories: comments about text comprehension (e.g., "I don't get this"), critiques of the text (e.g., "This smells like hyperbole"), and emotional responses to the text (e.g., "Tsk, I disapprove.") In addition to these prefabricated sentiments, readers select from a set of themes defined by the teacher, tying what students are reading to specific concepts from the curriculum. Readers' comments are aggregated and displayed so that students can grasp what their classmates think about what they are reading and how their thoughts relate to the topics discussed in class.

Selected tools implicated in various types of social reading are compared in Tables 1a and 1b in terms of their features (e.g., word clouds, annotations, heat maps, etc.), copyright licenses (i.e., open or closed), costs (i.e., free or proprietary), and general purposes (i.e., close reading, commentary aggregation, or textbook highlighting). In addition, the origins of the tools are indicated (i.e., internal university grants, external research grants, or commercial initiatives).

---

[1] Research reports on the use of Ponder in secondary schools are available on the website's blog at http://blog.ponder.co/category/research/

Table 1a
Social Reading Systems

	eComma	Gdocs	eMargin	Highlighter
Format	Drupal Website module (plugin)	Website	Website	Website
Producer	COERLL, University of Texas at Austin	Google	RDUES, Birmingham City University	Highlighter
Sponsors	U.S. Dept. of Education; University of Texas at Austin	Google	Birmingham City University	Investors Founders Co-Op, etc.
Cost	Free	Free	Free	No fee for publishing, but end users must purchase the document to annotate the text.
Hosting	Self Hosted Drupal module	Hosted on Google server	Hosted on eMargin server	Hosted on Highlighter server
Open License	YES	NO	NO	NO
Download/ Customizable source code	YES	NO	NO	NO
Support for teams, groups, or classrooms	Not built in, but several Drupal modules can add that functionality. COERLL uses the Organic Groups Drupal modules.	YES (use of folders and sharing options)	YES	YES
Original purpose	Textual analysis	Text editing tool	Textual analysis	Study tool
Features	• Tag/comment cloud • Word cloud of text • Heatmap of annotations • Reply on comments • Annotation text needs to be saved as Drupal node before it can be annotated. • Unstable <IE 8 browser support • Uses Drupal panels • Custom database queries (Drupal views) • Additional features: See Drupal modules	• Reply on comments • Annotations displayed in margin • Annotations are color coded • Document importer (PDF, Word, ...) • Works in all browsers, native to Chrome	• Annotations are displayed in popup box and in margin • Annotations are color coded • Dictionary search field integrated • Sharable links to specific annotations	Document importer (PDF, Word, ...)
FERPA friendly	YES	YES	YES	YES
Analytics	NO	NO	NO	YES

Table 1b
Social Reading Systems (continued)

	Marginalia	Annotator	Ponder	Classroom Salon
Format	Self hosted software and Moodle Website plugin	jQuery plugin/ API and WordPress Website plugin	Browser plugin and iPhone/iPad app and Website	Browser plugin and iPhone/iPad app and Website
Producer	Geof Glass, Simon Fraser University; U.S. Department of Education	Open Knowledge Foundation	Parlor Labs Inc.	Carnegie Mellon University
Sponsors	U.S. Department of Education; Simon Fraser University	Open Knowledge Foundation	Parlor Labs Inc.	Carnegie Mellon University
Cost	Free	Free	For a limited time, free for instructor's first semester.	Free
Hosting	Self Hosted	Self Hosted jQuery plugin	Hosted on Ponder server	Hosted on Salon server
Open License	YES	YES	NO	NO
Download / Customizable source code	YES	NO	NO	NO
Support for teams, groups, or classrooms	NO	NO	YES	YES
Original purpose	Textual analysis	Textual analysis	Textual analysis	Textual and multimedia analysis
Features	• Overlaps of annotations are revealed on hover • Annotations are displayed in margin	• Heatmap of annotations • Unstable in <IE 8 browser support	• Annotation tools plugin can be used in whatever browser • Data reports for instructors and students • Engagement Analytics day-to-day to drive student's reading achievements and help teacher identify problems • Tags are set up for classes by instructors • Students identify "flag" excerpts from their readings with a predefined sentiment tag	• All text files, youtube, and vimeo videos can be uploaded and annotated • iPad and smart phone support • Salon analytics can export student data • Can make copies of Salon texts/files • Different view modes: Students can only see their own or all comments • Salon navigator shows similarly to a heat map hotspots of discussions • A summary view of all annotation activities and hotspots
FERPA friendly	YES	YES	YES	YES
Analytics	NO	NO	YES	YES

4. Research on Collaborative Annotation and Digital Social Reading

Research on the reading of print texts has already proven the value of note-taking and textual markup for comprehension (Bangert-Drowns, Hurley, & Wilkinson, 2004; DiVesta & Gray, 1972; Einstein, Morris, & Smith, 1985; Wilson, 1999). As students shift their literacy preferences from print to digital, researchers have begun exploring the effects of digital annotation on reading. Johnson, Archibalk, and Tenenbaum (2010) reported that community college students were able to improve their reading comprehension and metacognition when asked to compare their textual annotations with their teacher's annotations. Mendenhall, Kim, and Johnson (2011) explored the use of digital annotations by students enrolled in an L1 university English course. Mendenhall et al. claimed that collaborative annotations promoted a lively discussion among readers that, in turn, led to a deeper interpretation of the text. In similar fashion, Gao (2013) studied the use of digital annotation by preservice teachers in an educational technology course. Gao claimed that the use of digital annotation led to increased participation in reading discussions.

Similar to textual markup, classroom-based L1 collaborative reading has been empirically proven in a variety of educational settings (Vaughn & Klingner, 1999; Vaughn, Klingner & Bryant, 2001; Zoghi, Mustapha, Maasum, & Mohd, 2010). Classroom collaborative reading is said to improve comprehension by turning reading into a small-group activity during which classmates help each other to apply basic reading strategies, namely predicting content, identifying problems, getting the gist, and synthesizing textual parts into a whole. This form of collaborative reading—like all forms of collaborative learning—is grounded in a view of learning in which social interaction is taken to be essential for creating meaning and knowledge (Vygotsky, 1978). The zone of proximal development (ZPD) is a key construct of sociocultural theory, the Vygotskyan approach to social learning. The ZPD is commonly defined as the distance between what a learner can accomplish on her own as opposed to what she can accomplish with the help of more experienced peers or experts. According to Vygotsky (as cited in Lantolf, 2011), the ZPD "is the activity in which instruction (i.e., socialization at home and formal teaching at school) and development 'are interrelated from the child's very first day of life'" (p. 29).

Encouraged by studies of face-to-face collaborative reading in the classroom, researchers have begun to study the pedagogical efficacy of online discussion. In a year-long study of EFL university students, Chiang (2007) demonstrated that participation in "virtual literature circles" led to increased reading comprehension. Chiang noted that even students unable to fully participate in the online discussion due to limited communicative proficiency showed gains in textual understanding. In a similar study of EFL students, Murphy (2010) analyzed students' online discussions about literary texts and concluded, much like Chiang, that dialogic interaction increased reading comprehension. Kiili, Laurinen, Marttunen, & Leu (2012) investigated Finnish high school students who participated in collaborative online reading. Based on a close analysis of transcripts, they found that students adopted several effective strategies during online discussions: gathering and synthesizing information, asking questions, expressing opinions, agreeing/disagreeing with arguments, making inferences, and proposing solutions. Coffey (2012) claims that students profit from discussion transcripts for their own personal reflection and analysis. In addition, Coffey argues that technology-enhanced reading discussion allows students to enlarge their social network beyond their classmates and to connect with a wider variety of readers.

Researchers are beginning to explore the effects of digital social reading in the L2 context. Three studies conducted at Taiwanese universities focused on different aspects of digital social reading in L2 classrooms such as the use of annotation for guiding discussion, the optimum size of a collaborative reading group, and student awareness of rhetorical text structure. In a controlled study of 94 EFL students, Yang, Zhang, Su, & Tsai (2011) tested the effects of a multimedia web-based forum combined with a collaborative annotation tool compared to other forms of digital communication lacking an annotation component (e.g., e-mail and instant messaging). Yang et al. defined annotation as "an explicit expression of knowledge that is attached to a document to reveal the conceptual meanings of an annotator's implicit thoughts" (p. 45). The experimental annotation group and the control group were determined to have the same reading proficiency in English based on reading comprehension tests. The study required the two groups to read five different English texts throughout the semester and to discuss the texts with their groups. The results showed that the experimental group outperformed the control group based on reading comprehension tests administered after each treatment. Yang et al. suggested that the results indicate that web-based collaborative annotation is a successful example of computer-supported collaborative learning.

Chang and Hsu (2011) tested the use of collaborative annotation and translation in an EFL course. Delivered via a mobile device, the annotation tool was developed to support a digital social reading environment. The goal of the study was to determine the optimum number of users for a social reading group. Experimental results indicated that groups of two, three, and four students had significantly higher levels of reading comprehension than groups of one student or groups of five students. In addition to examining the effects of group size on collaborative reading outcomes, the study also explored student perception of the tool. Attitudinal surveys indicated that 75% of the students felt that the system was easy to use, and 66% of the students felt satisfied with the tool.

Finally, Lo, Yeh, and Sung (2013) examined the effects of Paragraph Annotator (an annotation system) on EFL readers. This tool allows readers to analyze the paragraph structure of webpages by annotating paragraph elements such as topic sentences and supporting details. In addition, users can add personal comments to an annotated paragraph element. The developers compared an experimental group of EFL students who used Paragraph Annotator while reading selected webpages to a control group who took notes off line while reading the same webpages. The results revealed that students using Paragraph Annotator while reading English texts online scored significantly higher on cued recall tests and free recall tests. In addition, participants using Paragraph Annotator responded positively to perceived usefulness and ease of use.

5. Case Studies of eComma

Since 2010, the Center for Open Educational Resources and Language Learning has been pilot testing eComma in L1 and L2 classrooms at universities throughout the United States (Blyth, 2013). The goal of the pilot testing is to gather usability information as part of ongoing formative assessment in order to improve the tool's design and functionality. In addition, the developers are interested in determining whether use of the tool changes teachers' understanding of L2 literacy instruction. As part of the pilot testing, instructors were asked to write up their use of eComma in a case study format in order to provide answers to several questions, including the following:

1. What prompted you to use eComma?
2. How did you embed the tool into your reading lesson?
3. What was the actual assignment?
4. How did your students perform?
5. What did you learn from using the tool?

The case studies presented below were chosen to represent a range of perceived affordances according to different L2 environments, including both undergraduate and graduate courses. Block quotations are used without quotation marks, except when quotation marks were included in the original feedback from users of eComma.

5.1 First Year French Course (Lecturer, Private East Coast University)

As a teacher of first and second-year French, I have long been frustrated by approaches to L2 reading. Two assumptions seem to prevail: 1) language content should center on concrete subjects and literal meanings, and 2) the use of the L1 should be avoided. I find both assumptions questionable because language is inherently metaphorical and students necessarily filter their L2 interpretations through their L1 subjectivities. So, I decided to create a social reading activity for my first semester French class that permitted the use of English as part of my students' interpretive process.

I compiled a selection of prose poems from Dany Laferrière's *Chronique de la dérive douce*, a book that recounts in 365 prose poems the first year (1976) that Laferrière spent in Canada as a refugee from the Duvalier regime in Haiti. I provided students with some brief background information on Lafferière as well as context-specific glosses for some of the new vocabulary. Using *eComma*, my students noted their expectations, read the text, and shared their comments with each other in English. The entire assignment was done asynchronously as homework.

Because I was interested in understanding my students' meaning-making activity, I focused on *eComma's* commenting function. I was pleased to find evidence of a wide range of interpretive strategies in their comments:

- Evaluating the Meanings of Foreign Words

"We are limited, here, by our ability to work with a word in French while not knowing *exactly* what it translates to in French. Does 'interminable' [have] the exact same [meaning] in French as it does in English? If so than 'Un baiser interminable' carries a negative connotation: a kiss that's unbearably, almost annoyingly long and dragged out. But, in French, 'interminable' may also translate to other, less loaded words, such as perhaps 'endless.' Then, connotation changes: perhaps the author isn't put off by the extravagance of the kiss but, rather, admires it?"

- Reflecting Upon Cultural Differences

"It also makes me think of the huge cultural divide between Montreal and Haiti despite the fact that French is spoken in both locations."

- Interpreting the Meanings of Textual Features

"I am sensing a shift in this stanza. In the beginning, he is referring to a dictatorship, which we can identify as being evil and strict.... Then, the tone shifts when he refers to himself as being a virgin, which we can identify as being innocent and pure ... I believe that this shift is meant to shock us and create an uncomfortable tone that captivates us and encourages us to read more."

- Connecting Reading to Personal Experience

"Little ironic that we're undertaking the same 'voyage' in having to say everything (or, at least, having to try to say everything) in a language that is not our mother tongue, i.e. French class!"

- Co-constructing Meaning

"I actually did not get that impression when I read it, but it really does make sense and I completely agree with you.... It's as if he can make a physical connection in this new world, but he cannot make a deeper one because so many differences exist ..."

My Preliminary Conclusions
- The use of English allows students to take greater risks in expressing their interpretations.
- Social reading permits students to exploit the text on many levels and thus precludes certain pre-reading activities.
- Social reading models effective reading strategies for learners who may not employ them otherwise.
- The inductive nature of social reading (discovering meaning through social interaction) heightens the appreciation of the text, of reading in a foreign language, and of the fruitfulness of collaboration.

Social reading is a new form of reading that requires a reexamination and adjustment of current L2 pedagogy. I would like to try a similar activity for intermediate courses to explore the effect of the use of the L1 and the L2. In particular, I would like to give different classes the same text and ask one class to comment in English, a second class to comment in French, and a third class to comment in either language.

5.2 Second Year French Course (French professor, Public Midwestern University)

Despite my efforts to simplify the readings in my fourth semester French course, my students often complain that they spend hours hunting for definitions in the dictionary. I attempted to solve this problem with *LitGloss*, a project from the University of Buffalo that provides foreign language texts with grammatical and cultural glosses. And yet, I discovered that glosses were no panacea. In fact, it seemed that glosses actually made my students more passive and less apt to struggle with the demands of close reading. My hope was that *eComma* would enable my students to "crowd source" their reading burdens. I also felt convinced that *eComma*'s hypermedia environment would heighten my students' awareness of the process of textual interpretation.

The readings in the course represent different literary genres. To contextualize the readings, the textbook gives background information about the author and the relevant artistic movement. For example, in its introduction to the poem *Liberté*, the textbook describes the author, Paul Eluard, as one of the major figures in the surrealist movement. In addition to the work of Eluard, the textbook briefly reviews the work of other surrealist artists, including the Belgian painter Magritte whose famous autoportrait pictures a green apple floating in front of a man dressed in an overcoat and bowler hat.

After viewing several surrealist paintings together, I instructed my students to not only "read" the poem on *eComma*, but to annotate the text's surreal non sequiturs and juxtapositions. I told my students that the assignment was to be done asynchronously in either English or French. Finally, I gave the students a deadline that left me enough time to review their work before the following class.

My students' annotations and comments exceeded my expectations. Everyone posted multiple comments and many students replied to comments. In other words, the students were reading each other reading the text. The result was a collaborative commentary that was more nuanced and creative than any I had ever received from a student at this level. Of course, this wasn't the product of a single student, but of the crowd.

I used several *eComma* features to guide the in-class discussion. I found the heat maps useful for pointing out that the most annotated stanzas were the first and the last. This led to an interesting discussion about the role of the first stanza in setting up reader expectations and the role of the final stanza in corroborating those expectations. I used the sorting feature to display comments of individual users. Thanks to this feature, I was able to demonstrate that some students paid more attention to concrete images while others attended to abstractions. Again, this insight became fodder for an interesting discussion about reading styles. And finally, my students did something that was completely unexpected but very much in keeping with the activity. Several students annotated the text with visual images taken from surrealist paintings. It was a great way to extend the previous day's discussion of the interaction between surrealist literature and the visual arts.

In the future, I envision a multi-step reading task that begins with a focus on forms followed by a focus on cultural frames of meaning. I witnessed an English literature professor who began an eComma lesson by giving each student a note card. On each card was written a different rhetorical device—metaphor, metynomy, repetition, etc. The instructor gave the students 5-10 minutes to read through the text and tag examples of the rhetorical device on their note card. At the end of this "pre-reading" activity, the instructor quickly reviewed the devices with the students. Next, the instructor told the class to begin their actual reading of the text. This lesson gave me ideas for doing something similar by starting class off with a brief form-focused activity that would result in a highly glossed text that would facilitate reading comprehension and serve as a quick grammar review.

5.3 Applied Linguistics Graduate Course (Applied Linguistics Professor, Public Southwestern University)

I used *eComma* in a graduate level seminar titled *Literacy through Literature*. The course explores the role that literary texts and aesthetic reading play in the development of second language literacies. Literacy is used here both in the traditional sense of the reading of printed texts and in the wider sense of multiliteracies, which include social and cultural literacies as well as new media literacies.

One of the readings for the course was a chapter from narratologist Marie Laure Ryan's book *Narrative as Virtual Reality* (2001) on immersivity and interactivity as two modes of reading narrative texts. At the end of this chapter, Ryan introduces an example from the book *If on a Winter's Night a Traveler by* Italo Calvino. I decided to use *eComma* to allow students to consider Ryan's distinction between immersive and interactive reading, and to examine what kinds of pedagogical interventions might encourage one or the other mode of engagement. I instructed my students to read Calvino's text on *eComma* during the class period and to provide a minimum of 3 annotations.

After the first few students had created annotations, I noticed that almost everything that had been posted was an experiential response. I added a couple of comments of my own, including one about Calvino's use of the second person. I was curious to see if students would follow my lead or if they would continue to post about their personal reading experiences. A couple of people did respond to my comment, such as one student who commented:

For things that "talk to us" in the way that this text does, several things come to mind: 1) 2nd person narration is sometimes used in guided relaxation/meditation/calming exercises. People with a hard time relaxing might close their eyes and play a CD with something calming to focus their attention, like "You are sitting in a calm meadow. You see a deer on the edge of the trees and it runs across the meadow" etc. Using a 1st or 3rd person narrator instead would not have quite the same effect, I think. 2) Just off of the top of my head, tv shows can employ "talking to us" as a means of creating a more personal connection with the viewer, bringing him/her more into the action (such a in Sesame Street, when the characters sometimes ask the viewers questions and then pause for a response). It can also be used for comical effect, although I can't think of a good example of this besides Arrested Development, which isn't really quite the same (the narrator talks to us, but I don't think the characters ever do).

This comment, and comments like it, yielded productive material for our in-class discussion of the degree to which a text encourages either immersive or interactive readings. While this student attempted to build upon what I had said by identifying other familiar literacy practices in which the second person was used, the better portion of the annotations pertained to their personal reading experiences. Many of these related to the particular experience of working with *eComma*.

I had found *eComma* easy to use, so I decided not to give an introduction to the software. For this reason, a couple of students had trouble discovering how to add annotations, but this in itself provided some fruitful fodder for conversation. For example, one student connected her frustrations with the unfamiliar technology with her responses to the narrator's second person address:

I agree with the technology thing – I cannot figure out how to make a new comment so I am only replying to others. I also feel an annoyance with the being told what to do parts, but not because I feel like I, personally, am being told what to do, but because I think the author is trying to be funny, and I feel this has been done before? Or is at least a bit cliché. But in the final paragraph, when he says that I am no longer a person who expects much from anything anymore (I also cannot figure out how to tag…) that's when I suddenly got drawn into the text. This is when I feel like I was being addressed, not because of the second person, but because he was describing a human emotion I could relate to. I started considering this a literary text when it stopped trying to sound like a funny intro to a Eng101 text on how to read and starting allowing me to interpret meaning. Perhaps this is my own definition of literature.

This quote sparked a discussion in class about the potential effects of pedagogical applications that might encourage immersive and interactive readings. At the end of our in-class discussion, I asked students whether they could envision working with *eComma* in their classes and what ideas they had for how to implement it. Even the students who had had problems using the software agreed that the software held a lot of potential for working with texts in an L2 context. They were most excited about the flexibility of the software. Several noted that in addition to the kind of general response assignment that they had been given, the program could also be used as a space to share students' findings during a text-anchored Internet research activity.

5.4 Sociolinguistics Graduate Course (Linguistics Professor, Public Midwestern University)

Every couple of years, I teach *Introduction to French Sociolinguistics*. One of the major topics of this course is the social meaning attached to variation in the French language. Over the years, I have developed activities to help my students become more aware of language variation. For example, I typically have my students watch videos of naturally occurring conversations between French speakers. With the help of a transcript, I ask them to identify as many examples of language variation as possible. Another exercise that I find useful is based on a short story written in Louisiana Creole. Since there is no standard way to write Louisiana Creole, the author invented his own spellings, borrowing heavily from French orthography. As one might imagine, the story is full of variation. For my eComma experiment, I uploaded a passage from *the short story* onto *eComma* so my students could share their work with each other. As usual, I told them to annotate as many variable spellings as they could find. I also showed them how to use *eComma*'s word cloud feature.

My graduate students had no trouble using *eComma*'s annotation tools to mark up their texts. More importantly, they used *eComma*'s comment function to share their hypotheses with each other. Here is a good example of a hypothesis about the meaning of a group of related words:

To deal with the variation of *"mo/mo'/mò/mon,"* I pulled out all of the instances and grouped them together by usage. The first thing revealed is that *"mon"* is only used as an object. Whereas *"mo/mo'/mò"* may be used as subject pronouns. For example, the subject pronouns *"mo"* and *"mo'"* both appear in front of the verb *"cônain."* I have a feeling that the alternate spellings reflect that there is some (free) variation in the pronunciation of the word.

Overall, I was pleased with the results of the activity. I think that *eComma* can be used for consciousness-raising activities to help students discover form-meaning pairings. An example of this type of activity would be to have students annotate different past tense forms, e.g., preterit vs imperfect. Seeing the highlighted verb forms in context would help students develop their own hypotheses about the meanings of past tenses. The same could be done for other formal categories: singular vs plural, definite vs indefinite, active vs passive, indicative vs subjunctive.

6. Summary of Perceived Affordances

Despite salient differences, these four case studies illustrate several shared perceptions of the affordances of digital social reading as instantiated in the web-based application eComma.

6.1 Creating a ZPD for Less Expert Readers

Teachers often mentioned the pedagogical potential of turning an individual activity into a crowd-sourced activity during which readers who encounter difficulties seek assistance from others. For example, the teacher of beginning French felt that the tool allowed students to model effective reading strategies for each other. Similarly, in the sociolinguistics lesson, the teacher understood the tool as allowing students to model their strategies for constructing hypotheses based on data. Finally, the instructor of intermediate French was motivated by a desire to help students deal with the frustration of having to look up words in the dictionary. It appears that the instructor thought that students would prefer to ask each other for word meanings and that such a procedure would be faster and more efficacious than "hunting for definitions in the dictionary." Each of these uses of eComma demonstrates how group activity is perceived by teachers as creating a coconstructed scaffold to assist students while reading.

6.2 Distributing the Cognitive Load

Teachers envision eComma as not only a way to make reading a collaborative activity, but also a distributed one. In other words, in traditional approaches to reading, different students are expected to perform the same activity, that is, to read a given text in the same manner. However, in the context of eComma, different students were allowed to make different contributions to the group. For example, the instructor of intermediate French cites an English literature lesson that begins with a distributed activity during which students tag the text for different rhetorical devices. It appears that the French instructor sees this as a profitable way for students to divide and conquer a text by each student contributing different but valuable elements of meaning to the group commentary. In brief, some of the teachers who used eComma believed that when reading becomes a group activity, individual readers are free to read in different ways.

6.3 Synthesizing Several Activities into a Single Activity

Teachers reported that eComma allowed them to synthesize the traditional sequence of reading activities (i.e., prereading, reading, and postreading) into one activity (i.e., reading). In many L2 contexts, prereading activities serve as advance organizers, the purpose of which is to prime certain meanings by activating students' cognitive schemata. The basic idea is that students will have an easier time comprehending the text when some of the meaning elements can be anticipated. According to the lecturer of first-year French, prereading activities seem to be superfluous when students have access to eComma's suite of tools that allow them to explore textual meaning from the beginning. Rather than relying on prereading guidance, the applied linguistics professor actually took part in the reading activity with the students. By reading alongside the students, this teacher was able to guide the reading process in a moment-by-moment fashion.

6.4 Aggregating Behavior

Because eComma records and displays interpretive behavior such as annotations and comments, students and teachers can analyze how texts are interpreted at the group level. This allows teachers to discuss reading behavior in terms of the longitudinal process of an interpretive community, that is, a patterned way of interpreting a text. Several teachers noted that postreading discussions were prompted or guided by the use of eComma's features. The teacher of intermediate French used the heat map feature to point out which parts of the text received the most attention as represented by aggregated annotations. Moreover, this teacher also used the participant list feature to demonstrate how different students focused their attention on different parts of the text.

6.5 Blending Different Types of Digital Reading

Finally, many teachers noted that reading with eComma was a highly computer-mediated enterprise. In addition to using the suite of tools that eComma provided, students routinely used other hypermedia tools to create meaning such as Google Earth or Wikipedia or Word Reference. It appears that the students made heavy use of hyper reading tools yet rarely used machine reading tools such as the folksonomies (folk taxonomies) created by the program's comment cloud or tag cloud functions. However, the teachers used the machine reading tools to guide postreading discussion. It would seem that different texts and different assignments prompt different types of digital reading, as suggested by Hayles (2012).

7. Conclusion

This chapter has focused on an emergent form of "the discursive practice of critical interpretation" (Allington, 2009), an online, collaborative discourse that results from digital social reading. Spurred by the rapid development of e-reading devices, digital social reading is clearly on the rise. As Hayles (2012) argues, this new form of digital literacy will present a considerable challenge for many language and literature teachers whose mindsets have been formed by print culture (Baron, 2013). For, as the teacher who used eComma in a first-year French class asserts, "social reading is a new form of reading that requires a reexamination and adjustment of current L2 pedagogy."

Given that social reading is still relatively young, it is too soon to predict how language teachers will incorporate collaborative annotation and networked critical interpretation into their classrooms. Reviews of the research of computer-mediated language learning demonstrate that teachers invariably understand new technology in terms of familiar practices, an example of the new-wine-in-old-bottles phenomenon (Blyth, 2008; Chun, 2008). Nevertheless, the teachers who used eComma were quick to grasp its innovative affordances. The case studies suggested that teachers did not view digital social reading as a way to do the same thing better, but rather as a way to do new and different things, such as reading a text *with* one's students, analyzing group patterns of interpretation, and marking up a text with multimedia, multilingual glosses.

According to Jenkins (2006, 2009), what makes such digital practices truly innovative is their social and collaborative nature. In fact, Jenkins (2009) uses the term "participatory culture" as a new framework for understanding digital literacy practices (p. 1). Jenkins outlines five defining features of online participatory culture: relatively low barriers to artistic expression and civic engagement, support for creating and sharing one's projects, informal mentorship, a belief that contributions matter, and a sense of social connection (2009, p. 1). As Jenkins puts it, "participatory culture shifts the focus of literacy from one of individual expression to community involvement. The new literacies almost all involve social skills developed through collaboration and networking" (2009, p. 1).

Reframing digital literacy as participatory culture has important entailments. The most consequential of which is that digital practices are no longer viewed as replacing print practices, but rather as transforming individual practices into social ones. This is to say that print culture is not being replaced by digital culture as many teachers may think; instead, literacy culture as a whole is becoming more participatory. The effect of this reframing can be understood by paraphrasing a statement by Maryanne Wolf, a well known researcher who studies the "reading brain." Wolf frames her research in these terms:

> What would be lost to us if we replaced the skills honed by the reading brain with those now being formed in our new generation of "digital natives," who sit and read transfixed before a screen? (2007, p. 221)

Digital social reading and participatory culture affords teachers and researchers to reframe the question as follows:

> What would be *gained* if we *bridged* the skills honed by the *print* reading brain with those now being formed in our new generation of "digital natives," who sit and read *(with others)* before a screen?

References

Allington, D. (2009). Private experience, textual analysis, and institutional authority: The discursive practice of critical interpretation and its enactment in literary training. *Language and Literature, 21*, 211–225.

Annotator [Computer software]. Cambridge, England: Open Knowledge Foundation.

Anderson, S. (2011, March 4). What I really want is someone rolling around in the text. *The New York Times.* Retrieved from http://www.nytimes.com/2011/03/06/magazine/06Riff-t.html?pagewanted=1&_r=3&ref=magazine&src=me

Bangert-Drowns, R. L., Hurley, M., & Wilkinson, B. (2004). The effects of school-based writing-to-learn interventions on academic achievement: A meta-analysis. *Review of Educational Research, 74,* 29–58.

Baron, N. (2013). Reading in print or onscreen: Better, worse, or about the same? In D. Tannen & A. M. Trester (Eds.), *Discourse 2.0: Language and new media* (pp. 201–224). Washington, DC: Georgetown University Press.

Bauerlein, M. (2008). *The dumbest generation: How the digital age stupefies young Americans and jeopardizes our future.* New York, NY: Penguin.

Bauerlein, M. (Ed.). (2011). *The digital divide: Arguments for and against Facebook, Google, texting, and the age of social networking.* New York, NY: Penguin.

Blyth, C. (2008). Research perspectives on online discourse and foreign language learning. In S. Magnan (Ed.), *Mediating discourse online* (pp. 47–70). Amsterdam, The Netherlands: John Benjamins.

Blyth, C. (2013). eComma: An open source tool for collaborative L2 reading. In A. Beaven, A. Comas-Quinn, & B. Sawhill (Eds.), *Case studies of openness in the language classroom* (pp. 32–42). Dublin, Ireland: Research-publishing.net. Retrieved from http://research-publishing.net/publication/chapters/978-1-908416-10-0/Blyth_108.pdf

Bridle, J. (n.d.). *Open bookmarks* [Website]. Retrieved from http://www.openbookmarks.org/social-reading/

Carr, N. (2010). *The shallows: What the internet is doing to our brains.* New York, NY: Norton.

Chang, C.-K. & Hsu, C.-K. (2011). A mobile-assisted synchronously collaborative translation-annotation system for English as a foreign language reading comprehension. *Computer Assisted Language Learning, 24,* 155–180.

Chiang, M.-H. (2007). A novel idea: English as foreign language reading via virtual literature circles. *English Teaching and Learning, 31,* 367–403.

Chun, D. (2008). Computer-mediated discourse in instructed environments. In S. Magnan (Ed.), *Mediating discourse online* (pp. 15–45). Amsterdam, The Netherlands: John Benjamins.

Classroom Salon [Computer software]. Pittsburgh, PA: Carnegie Mellon University.

Coffey, G. (2012). Literacy and technology: Integrating technology with small group, peer-led discussion of literature. *International Electronic Journal of Elementary Education, 4,* 395–405.

DiVesta, F. J., & Gray, G. S. (1972). Listening and note taking. *Journal of Educational Psychology, 63,* 8–14.

eComma Drupal Module [Computer software]. Austin: Department of English/Center for Open Educational Resources and Language Learning, University of Texas at Austin.

Einstein, G. O., Morris, J., & Smith, S. (1985). Note-taking, individual differences, and memory for lecture information. *Journal of Educational Psychology, 77,* 522–532.

eMargin [Computer software]. Birmingham, England: Research and Development Unit for English Studies, Birmingham City University.

Gao, F. (2013). A case study of using a social annotation tool to support collaborative learning. *The Internet and Higher Education, 17,* 76–83.

Gibson, J. J. (1977). The theory of affordances. In R. E. Shaw & J. Bransford (Eds.), *Perceiving, acting, and knowing* (pp. 67–82). Hillsdale, NJ: Erlbaum.

Gibson, J. J. (1979). *The ecological approach to visual perception.* Boston, MA: Houghton Mifflin.

Hayles, K. (2012). *How we think: Digital media and contemporary technogenesis.* Chicago, IL: University of Chicago Press.

Howard, J. (2012, November 26). With "social reading," books become places to meet. *The Chronicle of Higher Education.* Retrieved from http://chronicle.com/article/Social-Reading-Projects/135908/

Jackson, H. J. (2001). *Marginalia: Readers writing in books.* New Haven, CT: Yale University Press.

Jenkins, H. (2006). *Convergence culture: Where old and new media collide.* New York, NY: New York University Press.

Jenkins, H. (2009). *Confronting the challenges of participatory culture: Media education for the 21st century.* Chicago, IL: MacArthur Foundation. Retrieved from http://www.macfound.org/media/article_pdfs/JENKINS_WHITE_PAPER.PDF

Johnson, T. E., Archibalk, T. N., & Tenenbaum, G. (2010). Individual and team annotation effects on students' comprehension, critical thinking, and meta-cognitive skills. *Computers in Human Behavior, 26,* 1496–1507.

Kiili, C., Laurinen, L., Marttunen, M., & Leu, D. J. (2012). Working on understanding during collaborative online reading. *Journal of Literacy Research, 44,* 448–483.

Lantolf, J. (2011). The sociocultural approach to second language acquisition. In D. Atkinson (Ed.), *Alternative approaches to second language acquisition,* (pp. 24–47). London, England: Routledge.

Laurence, D. (2008). Learning to read. *ADE Bulletin, 145,* 3–7.

Lo, J., Yeh, S-H., & Sung, C.-S. (2013). Learning paragraph structures with online annotations: An interactive approach to enhancing EFL reading comprehension. *System, 41,* 413–427.

Luks, J. (2013). *Le littéraire dans le quotidien: Resources for a transdisciplinary approach to reading/writing at the first and second year levels of college French.* Austin: Center for Open Educational Resources and Language Learning, University of Texas at Austin. Retrieved from https://drive.google.com/folderview?id=0Byg7PyauMJRScGxPdlpReVJQaUU&usp=sharing

Mendenhall, A., Kim, C., & Johnson, T. E. (2011). Implementation of an online social annotation tool in a college English course. In D. Ifenthaler, P. Isaias, J. M. Spector, P. Kinshuk, & D. Sampson (Eds.), *Multiple perspectives on problem solving and learning in the digital age* (pp. 313–323). New York, NY: Springer.

Murphy, P. (2010). Web-based collaborative reading exercises for learners in remote locations: The effects of computer-mediated feedback and interaction via computer-mediated communication. *ReCALL, 22,* 112–134.

Neisser, U. (1987). From direct perception to conceptual structure. In U. Neisser (Ed.), *Concepts and conceptual development: Ecological and intellectual factors in categorization* (pp. 11–24). Cambridge, England: Cambridge University Press.

Ong, W. (1982). *Orality and literacy.* London, England: Routledge.

Pettitt, T. (2007). *Before the Gutenberg parenthesis: Elizabethan-American compatibilities.* Paper presented at Media in Transition: Creativity, Ownership and Collaboration in the Digital Age, Massachusetts Institute of Technology, Cambridge. Retrieved from http://web.mit.edu/comm-forum/mit5/papers/pettitt_plenary_gutenberg.pdf

Thorne, S., & Reinhardt, J. (2008). "Bridging activities," new media literacies, and advanced language proficiency. *CALICO Journal, 25,* 558–572.

van Lier, L. (2004). *The ecology and semiotics of language learning: A sociocultural perspective.* Norwell, MA: Kluwer.

Vaughn, S., & Klingner, J. K. (1999). Teaching reading comprehension through collaborative strategic reading. *Intervention in School and Clinic, 34,* 284–292.

Vaughn, S., Klingner, J. K., & Bryant, D. (2001). Collaborative strategic reading as a means to enhance peer-mediated instruction for reading comprehension and content-area learning. *Remedial and Special Education, 22,* 66–74.

Vygotsky, L. S. (1978). *Mind in society: The development of higher psychological processes.* Cambridge, MA: Harvard University Press.

Wilson, K. (1999). Note-taking in the academic writing process of nonnative speaker students: Is it important as a process or a product? *Journal of College Reading and Learning, 29,* 166–79.

Wolf, M. (2007). *Proust and the squid: The story and science of the reading brain.* New York, NY: Harper.

Yang, S., Zhang, J., Su, A., & J. Tsai (2011). A collaborative multimedia annotation tool for enhancing knowledge sharing. *Interactive Learning Environments, 19,* 45–62.

Zoghi, M., Mustapha, R., Maasum, T., & Mohd, N. R. (2010). Collaborative strategic reading with university EFL learners. *Journal of College Reading and Learning, 41,* 67–94.

Chapter 10

Digital Literacy in Academic Language Learning Contexts: Developing Information-Seeking Competence

MAARIT MUTTA
University of Turku (Finland)

SANNA PELTTARI
University of Turku (Finland)

LEENA SALMI
University of Turku (Finland)

ALINE CHEVALIER
University of Toulouse II-Le Mirail (France)

MARJUT JOHANSSON
University of Turku (Finland)

Abstract

This chapter reports on two studies. The first study (Study A) concentrates on information seeking in one's first language (L1, French and Finnish university students) and in a foreign language (French as L2; Finnish university students). The second study (Study B) concentrates on the effects of raising awareness in information seeking in L2 (French and Spanish). The chapter analyzes the use of sources, keywords, search strategies, and the development of information-seeking competence as a part of digital literacy in the academic context. Digital literacy refers to the critical assessment and use of digital technologies and the competencies in digital communication and discourse. The results from the first study indicate that the native speakers of French needed less time than L2 learners of French when searching for information. However, the L2 learners searching for information in Finnish (L1) formulated more queries and used more keywords than the native speakers or those searching in L2 (French). When the two sets of data were compared, searching for information using online search engines was evidently the most widely used information-seeking strategy. However, the students developed several individual seeking paths and strategies to complete the tasks. For example, the students used different languages to find the appropriate answer or translation equivalent. The chapter provides concrete examples to illustrate different information-seeking paths.

1. Introduction

Information seeking[1] on the internet has become a fundamental activity in contemporary society (Singer, Norbisrath, & Lewandowski, 2013), at least in high-income countries where young people are characterized as *digital natives* (see, for instance, Barton & Lee, 2013). This entails the discussion of the relevant and critical use of the material found on the internet in terms of literacy (Gutiérrez & Tyner, 2012; Olson & Torrance, 2009). Scholars from many different fields, for example, the information and educational sciences, discuss the phenomenon with a large range of concepts that can cause confusion and misunderstandings. Gilster (1997) used the term *digital literacy*, but many other terms have been introduced, for example, *multiliteracies* (The New London Group, 1996), *web literacy* (Mackey & Ho, 2005), *information literacy* (Mackey & Ho, 2005; Pinto, 2010; Pinto & Sales, 2010), *media and information literacy* (UNESCO, 2008), and *media literacy and digital competence* (Gutiérrez & Tyner, 2012). In our approach, we start in the field of digital communication and discourse, which will serve as the basis for our definition of digital literacy and lead to our pedagogical view of language learning and teaching.

This new type of literacy, or rather, literacies, requires new types of social agency that are complex (Cope & Kalantzis, 2010) because it not only covers the use of the internet but all screen-based media. The texts and discourses are multimodal and hypertextual, including technological affordances, namely technological actions that are embedded in the pages for users to perform (see, e.g., Kress, 2003). Furthermore, the multimodal representation on the internet increases conceptual-cognitive complexity (Kress, 2003). We understand *digital literacy* to comprise two dimensions (see Johansson, 2014). The first dimension covers the relevant use of digital technologies that includes critical assessment concerning the selection and use of software. To perform these activities, users need to analyze and understand the affordances of the search engines and sources they use to accomplish their tasks. The second dimension refers to the competencies in digital communication and discourse that are language-related or other semiotic competencies of digital texts (reading, writing, speaking, listening, and viewing, etc.) (Dervin, Johansson, & Mutta, 2007; Guitert & Romeu, 2009; Martin, 2005; Newrly & Veugelers, 2009; Thurlow & Mroczek, 2011). Users seek, share, distribute, and produce new knowledge on the internet in relevant and appropriate manners for their information and knowledge needs. Furthermore, digital literacy is not a static state; it is subject to change and is better described as an ongoing process depending on an individual's life situation. In this context, remaining digitally literate means to learn and to use new technologies continuously and to adopt all these processes in everyday life (Martin, 2005; Newrly & Veugelers, 2009).

---

[1] The term *information seeking* was chosen to be used in this chapter at the general level. Other possible terms were *information searching*, used here when related to task complexity, *information retrieval* (IR), and *data retrieval* (DR), with the last two being more common in the domain of information science studies.

Recently, interactivity has become an increasingly essential part of information seeking. New software has been designed to facilitate collaboration and sharing of information online, and therefore, there has been an enormous accumulation of distributed or collaborative knowledge (Kress, 2003; Dobson & Willinsky, 2009). This has changed users' relationships to knowledge and how they read and write. For instance, trust formation or quality assessment in digital information sources have been studied in the case of Wikipedia. The findings suggest that, in trust formation, users rely on a range of factors such as authorship, references, expert recommendation, quality of writing and editing, and verification via links to external reference sources (Rowley & Johnson, 2013). However, it appears that quality is also partly a subjective concept that depends on the user's unique point of view (Yaari, Baruchson-Arbib, & Bar-Ilan, 2011; see also Dobson & Willinsky, 2009).

Search engines, such as Google, and online materials are widely used for finding information quickly (Massey & Ehrensberger-Dow, 2011; Salmi & Chevalier, 2010). The use of a search engine requires the individuals to generate keywords relevant to their query or need, evaluate the relevance of the results provided by the search engine, and then to select one or more webpages to visit (e.g., Al-Eroud, Al-Ramahi, Al-Kabi, Alsmadi, & Al-Shawakfa, 2011; see also the model developed by Sharit, Hernandez, Czaja, & Pirolli, 2008). If the search engine does not provide relevant elements, the information-seeking activity becomes more complex because individuals have to reformulate their query by adding and/or suppressing keywords (e.g., Dommes, Chevalier, & Lia, 2011). The process becomes even more complex when the individuals have to search in a nonnative language (Aula & Kellar, 2009).

In the current chapter, translating in another language is a central part of the information-seeking procedure. For all types of translators, from those translating literature to those localizing software, the web has become an essential tool. Translating is not only putting words found in a dictionary one after another on the page, but transferring the ideas contained in a text in one language to another language—"an expert activity which constantly requires information" (Pinto & Sales, 2007, p. 532), and the transfer of ideas necessitates getting acquainted with the subject matter of the text. In both studies, translation and/or the multilingual information retrieval (Al-Eroud et al., 2011) are included in the tasks performed by the informants, but it is also a strategy used when seeking information. In regard to the tasks, they are normally divided into two main categories, namely information-seeking or information-search tasks, the latter requiring problem information, domain information and problem-solving information. Moreover, the task complexity can be objective or subjective (Singer et al., 2013).

Our study concerns native (L1) and nonnative (L2) academic learners who are majoring in foreign languages or other academic subjects. The academic context refers to studying in a certain scientific field and acquiring expertise in which general and field-specific information-seeking competencies are expected learning outcomes.

2. The Two Studies[2]

This chapter presents the results of two studies reporting one dimension of the information-seeking competence of digital literacy and how students use this critical skill as a part of their emerging academic expertise. More precisely, the objective is to study how students search for relevant information on the internet with a focus on their use of sources and keywords, search strategies, and changes in language use between the L1 and the L2. Information-seeking competence is studied in different academic learning contexts. This chapter has the following general research questions:

1. What type of seeking behaviors or search paths can be distinguished in different tasks by different groups? (Study A and Study B)
2. Do the L1 and/or L2 of the learners and/or the language in which the queries are made have an effect on seeking behaviors and strategies developed? (Study A and Study B)
3. Does the awareness raising have an impact on the search process? If yes, in which way? (Study B)

The first study concentrates on information seeking in one's L1 or L2, and the second study concentrates on the effects of raising awareness in information seeking. With the first set of data, native speakers of French ($N = 20$) were compared to nonnative speakers of French ($N = 30$), whereas with the second set of data, Finnish students of L2 French ($N = 10$) and Spanish ($N = 10$) participated in a study that consisted of a small-scale experiment. The studies are each presented in turn before comparing the results in Section 5, Strategies for Finding Information, and then discussing the results and drawing the final conclusions.

3. Study A: Information Seeking by Native and Nonnatives

3.1 Participants

In Study A, data from a total of 50 participants were analyzed. Native speakers of French, students ($N = 20$) recruited at a university in Paris, France, were compared to nonnative speakers of French, Finnish translator students ($N = 30$) recruited at a university in southern Finland. The Finnish students studied French Translation as their major or minor subject, with the exception of one participant, who studied the French Language as his major subject. The participants were recruited on a voluntary basis. The 20 native speakers of French originally took part in a study reported by Dommes et al. (2011). The 30 Finnish participants were divided in two subgroups so that one subgroup had the task of finding the information in French (L2), and the other in Finnish (L1). The following abbreviations are used to distinguish the three user groups: 20 French participants (FR-FR),

---

[2] The Finnish data were collected by Leena Salmi and Matti Nokelainen and then analyzed by Leena Salmi and Aline Chevalier in Study A. In Study B, Maarit Mutta and Sanna Pelttari collected and analyzed all the data. Marjut Johansson is responsible for the definition of digital literacy (Introduction, Conclusion)..

instructed to find information in French (16 women and 4 men, M_{age} = 21.1 years, SD_{age} = 3.26), 16 Finnish participants (FI-FR), instructed to find information in French (14 women and 2 men, M_{age} = 22.4 years, SD_{age} = 2.61), and 14 Finnish participants (FI-FI), instructed to find information in Finnish (13 women and 1 man, M_{age} = 22.1 years, SD_{age} = 2.48). Each participant had one hour and a half to complete the tasks and to provide background information.

3.2 Procedure and Method

The data were collected in an experimental session performed in a separate, quiet room using a tabletop computer. The sessions were recorded using screen capture software. All the participants were given nine questions that can be described as "factual tasks" (term used by Kim, 2009, p. 683) or "closed informational tasks" (term used by Aula, Khan, & Guan, 2010, p. 37), that is, a task where one has to find a definite answer to a closed question. The participants were asked to find the answer by using the Google search engine. Once they had found the answer on a webpage, they had to highlight it using the mouse.

The questions for all the participants were provided in French, on paper, question by question in a random order by the experimenter. Either one of the researchers or a research assistant acted as the experimenter. The questions varied in complexity, such that there were three complexity levels of questions. There were three simple questions, for example, "What is the height of the tower of Pisa?" There were three difficult questions, for example, "Where can bog rosemary be found in France?" In addition, there were three impossible questions, for which answers did not exist at all such as "What is the frequency of the radio station 'Radio Santé'?"

Since Study A concentrated on the information-seeking behavior and contained impossible questions, the participants were not asked to consider the reliability of the answer found; it was sufficient to find an answer. The Finnish participants working in L2 were also told that the test was not a language test, and they were encouraged to ask for the meaning of any of the words they did not understand before starting. However, not all of them asked questions in the first place, but if they appeared to hesitate, the experimenter ensured they had understood the contents of the question.

3.3 Results

We computed the following quantitative measures: number of correct answers (by complexity level; because no answer existed for the impossible questions, we only considered the simple and difficult ones), time used for searching for information (from the moment the participants started the search with Google until they highlighted the answer or declared abandoning to find the answer), number of formulations of the query, and the number of keywords used in formulating the query (the number of keywords that the question contained, and the number of keywords the question did not contain). These measures were submitted to an analysis of variance (ANOVA) with group (FR-FR versus FI-FR versus FI-FI) as between-subject factor and complexity of questions (simple versus difficult versus impossible) as within-subject factor. Post hoc tests were computed on significant effects (HSD Tukey test). The significance level was set at .05 for all statistical analyses. Partial η^2 was used as an index of the relative effect size. The descriptive analyses (means and standard deviations) are presented in Table 1.

Table 1
Means and (Standard Deviations) of Scores of Correct Answers, Search Times, Formulations of the Query and Keywords According to the Group of the Participants and the Complexity of the Questions

	FR-FR group			FI-FR group			FI-FI group		
	SQ	DQ	IQ	SQ	DQ	IQ	SQ	DQ	IQ
SCA	2.8 (.41)	.9 (.85)	—	2.75 (.58)	.63 (.72)	—	2.64 (.5)	.87 (.73)	—
ST	25.44 (15.22)	263.9 (104.68)	267.64 (82.95)	52.67 (23.14)	358.23 (104.67)	323.69 (97.2)	120 (44.33)	337.43 (154.55)	355.13 (140.36)
FQ	1.13 (.26)	4.17 (1.28)	4.28 (1.57)	1.38 (.38)	7.23 (2.21)	6.25 (2.11)	2.38 (.98)	7.62 (2.91)	8.64 (3.16)
KW	3.63 (.94)	5.92 (1.85)	5.65 (1.52)	3.63 (.59)	6.08 (1.57)	5.67 (.83)	3.74 (1.5)	7.36 (2.48)	7.48 (1.65)

Note. SQ = Simple Questions, DQ = Difficult Questions, IQ = Impossible Questions, SCA = Score of correct answers, ST = Search times (in seconds), FQ = Formulations of the query, KW = Keywords.

3.3.1 Number of correct answers

The group of the participants had no significant effect on the mean number of correct answers ($F(2,47) = 0.562$, *ns*). In contrast, the complexity level had a significant effect ($F(1,47) = 208.308, p < .0001, \eta_p^2 = .81$). The simple questions generated better performances than the difficult ones. The interaction between the group and complexity was not significant ($F(2,47) = 0.74$, *ns*).

3.3.2 Time used for searching

The group of the participants had a significant effect on the search time ($F(2, 47) = 7.362, p < .005, \eta_p^2 = .24$). Post hoc comparisons indicated that the FR-FR group needed less time than the FI-FR group ($p < .05$) and the FI-FI group ($p < .005$). There were no significant differences between the FI-FR and FI-FI groups. The complexity level also had a significant effect ($F(2,94) = 164.425, p < .0001, \eta_p^2 = .78$). As could be expected, post hoc comparisons indicated that the simple questions were solved faster than the difficult ones ($p < .0001$) and the impossible ones ($p < .0001$). There were no significant differences between the difficult and impossible questions. The interaction between the group and complexity was not significant ($F(2,94) = 1.205$, *ns*).

3.3.3 Formulations of the query

The group had a significant effect on the number of formulations ($F(2,47) = 24.369, p < .0001, \eta_p^2 = .51$). More precisely, all the Finnish participants (FI-FR and FI-FI) formulated more queries than the French participants (FR-FR; $p < .001$). There was no significant difference between the participants in the groups FI-FR and FI-FI.

The complexity of the questions also had a significant effect ($F(2,94) = 136.896, p < .0001, \eta_p^2 = .74$). The simple questions required fewer formulations than the difficult and impossible questions ($p < .0005$). The complexity × group interaction was significant ($F(4,94) = 5.862, p < .0005, \eta_p^2 = .20$). For the simple questions, there was no significant difference between the three groups. For the difficult questions, the number of formulations was higher for group FI-FI than for group FI-FR ($p < .005$) than for group FR-FR ($p < .05$). In contrast, for the impossible questions, only group FI-FI formulated queries more often than the group FR-FR ($p < .001$).

3.3.4 Keywords used

The group had a significant effect on the number of keywords used ($F(2,47) = 4.681, p < .05, \eta_p^2 = .17$). The participants who had to change language from French into Finnish (FI-FI) used more keywords than the two other groups ($p < .05$). The complexity of the questions also had a significant effect ($F(2,94) = 79.401, p < .0001, \eta_p^2 = .63$). The simple questions generated fewer keywords than the difficult and impossible questions ($p < .0005$). The complexity × group interaction was significant ($F(4,94) = 2.674, p < .05, \eta_p^2 = .10$). For the simple and difficult questions, no significant difference appeared between the three groups. In contrast, for the impossible questions, group FI-FI used more keywords than the two other groups ($p < .05$).

3.3.5 Search paths

In addition to the above quantitative measures, the videotapes were analyzed to identify information-searching strategies developed by the participants. As for the formulations used in the queries, the basic method for simple questions was to type keywords one after another in the Google text box, starting with the keywords issued from the question. In group FI-FI, the keywords were translated into Finnish. If that did not work, the participants limited the search by using quotes to group a part of the keywords as phrases, and some of them used more the advanced operators "+" and "-." The difficult questions were formulated in such a way that it was necessary to use a synonym to find the correct information. Contrary to the results reported by Aula et al. (2010 for unsuccessful tasks, the participants in Study A never formulated their queries as questions (i.e., phrases starting with a question word and ending with a question mark).

4. Study B: Awareness Raising

4.1 Participants

The study was conducted in a university in southern Finland. Young Finnish students of L2 French (Group A, $N = 10$) and Spanish (Group B, $N = 10$) took part in relevant translation courses and participated in a predominantly qualitative study that consisted of a three-part small-scale experiment: a translation task, a pedagogical intervention, and a translation commentary task. All students were recruited from the translation course. They participated voluntarily and received feedback on the tasks. The translation commentary task was evaluated only for the use of the experiment; it did not affect the course evaluation.

The French language group consisted of 8 women and two men ($M_{age} = 20.7$ years, age range: 19-24 years). Six of these participants were first-year students, 4 were at least third-year students, and 7 had French as a major and 3 as a minor subject. In this case, the major subject was English or Swedish. The Spanish group included 10 women ($M_{age} = 24.9$ years, age range: 19-35 years), and 6 of them were second-year students and 4 at least third-year students. In addition, 6 of the group had Spanish as a major and 4 as a minor subject. In this case, the major was French, Italian, or German.

4.2 Procedure and Method

The three-part experiment was conducted in the following manner. Figure 1 illustrates the steps of the procedure.

Figure 1
The Steps of the Three-Part Experiment

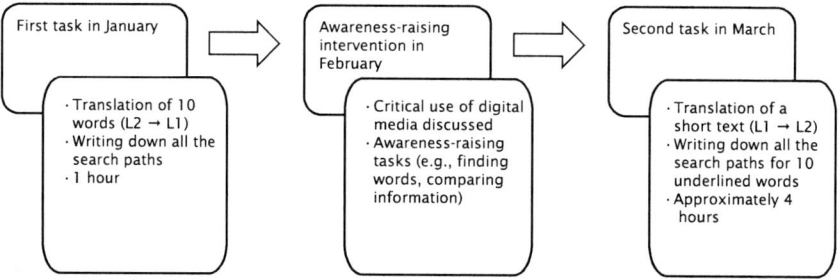

The first task was a translation task of 10 words from French or Spanish to Finnish during the course conducted in January. The time for this task was limited to one hour. The short instructions were given in Finnish and explained the task and the subject area of the words, which were partly different because of cultural differences but belonging to the same subject area (gastronomy, education, geography, animals or biology, culture, and economy). For example, in French, words such as *la galette des rois* 'king cake,' *le mémoire de master* 'master's thesis,' and *la calanque* 'a calanque' were used, and, in Spanish, words such as *la navaja* 'razor shell,' *el copago* 'copayment,' and *el colimbo chico* 'red-throated diver' were used. The participants were asked to write down all the search paths they made.

In February, before the second task conducted in March, teachers presented an awareness-raising intervention on the critical use of digital media. During the intervention, topics such as expertise in language, contemporary digital, multimodal and media literacy, and metacognitive awareness were discussed. The pedagogical intervention was based on the following studies: Chevalier and Tricot (2008), Dervin et al. (2007), Jewitt (2006), Johansson and Dervin (2007), Maroccia (2012), and Mangenot (2012). The intervention also contained awareness-raising tasks, such as finding words, comparing information found from different sources, and evaluating the trustworthiness of Wikipedia as a digital information source and the appropriateness of Google Translate as a tool in translation tasks.

The second task was a translation of a 168-word text from Finnish to French or Spanish with 10 words requiring further translation commentary. The same popular scientific text was used in both groups. The short instructions were given in Finnish and requested that the participants translate the whole text and write down all the search paths for the underlined words. By administering these three parts, it was determined which strategies and arguments were used by these learners in the process of searching and finding the relevant information being sought. In addition, the aim was to understand whether this type of processing approach could increase the awareness of digital literacy among students. The analysis of the students' information-seeking strategies and paths relied on earlier studies (Massey & Ehrensberger-Dow, 2011; Salmi & Chevalier, 2010), but, in addition, a grounded approach was used, which means that some categories emerged from the students' responses (see Woore, 2010).

4.3 Results

The queries made by students were categorized into six different search groups:

1. Google queries including all Google queries (e.g., translator, images, advanced) (G)
2. Wikipedia queries in all languages (WP)
3. MOT online dictionary searches in all languages (M)[3]
4. Other bi- or plurilingual dictionary queries (O)
5. Monolingual dictionary queries (ML)
6. Own implicit and or explicit knowledge with decision making (Own)

The last category refers to activities including the participants retrieving the word from their memory or deducing the meaning themselves without consulting a dictionary. It should be noted that all queries were calculated from explicitly indicated queries, even if they might include others. Example 1 illustrates the categorization as M + G (the text the participant wrote is translated from Finnish):

---

[3] MOT is a flexible language service developed by Kielikone Ltd. MOT features high-quality tools for everyday communication: dictionaries, proofreading software, and translation software for Windows, Internet and mobile platforms. MOT is widely used in Finnish universities.

(1) The aivosähkökäyrä [electroencephalography] translation from M "électroencéphalogramme *m* (EEG)" also provided a lot of results using the internet.

4.3.1 Quantitative differences between tasks and groups

The results indicate that there were an increased number of queries from Task 1 to Task 2 at the group level in Group A but not in Group B. Table 2 shows all the queries made by the participants in the two tasks.

Table 2
Number of Different Queries Made in Task 1 and Task 2

Group A			Task 1					Task 2					
P	S	G	M	WP	O	ML	Own	S	G	M	WP	O	ML	Own
1	16	8	0	3	5	0	0	28	4	10	4	1	9	0
2	22[a]	9	0	6	6	0	1	19	2	0	0	12	0	5
3	16[a]	12	0	4	0	0	0	37	19	3	2	13	0	0
4	28	6	4	7	10	0	1	34	21	0	3	10	0	0
5	21[a]	7	3	11	0	0	0	18	4	6	6	0	0	2
6	28[a]	17	0	9	1	0	1	14	3	2	0	8	0	1
7	11	4	0	0	7	0	0	19	4	0	3	12	0	0
8	15[a]	7	3	1	3	0	1	21	5	5	4	0	2	5
9	20[a]	0	4	5	11	0	0	53	27	5	4	17	0	0
10	8[a]	2	4	0	1	1	0	27	5	0	5	17	0	0
S	185							270						

Group B			Task 1					Task 2					
P	S	G	M	WP	O	ML	Own	S	G	M	WP	O	ML	Own
11	52	28	10	6	4	2	0	51	27	12	4	6	0	1
12	30	14	6	7	2	0	0	26	13	10	0	0	0	3
13	20	3	6	6	4	0	1	29	8	7	6	2	2	4
14	20	9	5	4	2	0	0	18	6	4	2	0	1	5
15	50	27	17	5	1	0	0	23	10	10	1	1	0	1
16	12	1	3	2	0	0	6	19	6	4	0	0	0	4
17	46	19	18	7	1	1	0	26	7	26	2	0	0	0
18	21	7	4	7	3	0	0	23	9	10	2	0	1	1
19	36[a]	10	12	2	8	3	0	36	19	10	2	0	2	2
20	30	10	7	3	9	0	0	27	11	10	0	4	0	2
S	317							278						

[a] indicates fewer than 10 words searched for. Only 3 participants had time to search for all the 10 words in Group A, 3 searched for 8 words, 2 searched for 7 words, 1 or 5 words, and 1 for four words (the last one due to technical problems). In Group B, however, only one participant did not have time to search for all the

10 words (9 words). Group A consisted of French students and Group B of Spanish students.

Note: P = participant, Σ = the total number of queries made, G = all Google searches, M = the online dictionary MOT, WP = Wikipedia, O = all other dictionaries, except monolingual dictionaries marked as ML, and Own = a participant's own implicit and/or explicit knowledge with decision making. All queries are calculated from explicitly indicated queries.

The number of queries was 185 in Task 1 (per person *meanquery* = 1.85, or 2.40 if only actual queries calculated) and 270 in Task 2 (per person *meanquery* = 2.70). The number of queries was 317 in Task 1 (per person *meanquery* = 3.17, or 3.20 if only actual queries calculated) and 278 in Task 2 (per person *meanquery* = 2.80). However, there was a great deal of individual variation in both groups.

The results indicate that in the whole data set (Groups A and B), the explicit queries from Google and Wikipedia decreased from Task 1 to Task 2 (39.8% vs. 38.4% and 18.9% vs. 9.1%, respectively). The use of other categorized sources increased from Task 1 to Task 2, namely MOT (21.1% vs. 24.5%), other dictionaries (15.5% vs. 18.8%), monolingual dictionaries (1.4% vs. 3.1%), and own knowledge (2.2% vs. 6.6%). However, there were some differences within each group (see Table 3).

Table 3
Percentages of Queries in Task I and Task II

	Group A		Group B	
	Task I	Task II	Task I	Task II
G	38.9	34.8	39.7	41.7
M	9.7	11.5	28.1	36.3
WP	24.9	11.5	16.4	6.8
O	23.8	33.3	8.2	5.4
ML	0.5	4.1	1.9	2.2
Own	2.2	4.8	2.2	8.3

Although the query results regarding different sources used are alike in the overall data, there were some differences between the groups. The results reveal that in Group B, unlike in Group A, the overall number of queries decreased in Task 2, but the number of Google queries slightly increased. The increased number of queries at the group level in Group A was most likely partly due to the difference of time on task being limited to 1 hour and some technical problems. In Task 2, the time on task was approximately 4 hours in both groups, varying from 2.5 to 6.5 and from 2.5 to 8 hours in French and Spanish, respectively. Some participants commented that the search time was most likely so long because, in addition to the translation task, they had to write down the search paths for the 10 underlined words. This could explain why some participants had fewer explicit queries in Task 2 than in Task 1 (participants 2, 5, 6, 12, 13, 14, and 16).

4.3.2 Successful searches and task complexity

As in some other studies (e.g., Singer et al., 2013), it is difficult to distinguish a successful from unsuccessful search in terms of time on task or number of queries. In Task 2, 38.5% of translations were erroneous in the whole data (77 out of 200 words). The number of errors varied from 0 to 8 (10 words per person) and the queries from 14 to 53. That is, the participants could make 21 or 51 queries with 2 errors in translation of the word in question or 26 queries with 1 or 6 erroneous answers. All in all, this result indicates that the correct answer did not correlate with the number of queries. However, there is a slight difference between the groups; Group A made 50 such errors (64.9%) and Group B only 27 (35.1%). One possible explanation for the differences between the groups may be the different ages and the academic year when the participants took the course. The students in Spanish group (B) took the course later in their studies and were older than those in French group (A) (M_{age} = 24.9 and 20.7 years, respectively). In addition, all students in the Spanish group (B) had completed at least two one-term-long Spanish translation courses before taking part in this particular course, whereas in the French group (A), the participants had completed at least one one-term-long translation course before this course. Moreover, students with French or Spanish as a minor subject might have completed equivalent courses in their main subject. However, there were no significant correlations between these factors.

It appears that the main reason for differences is individual search behaviors. The search tasks 1 and 2 were considered objectively quite complex. On the basis of the results related to the time spent on tasks 1 and 2 and success or lack of success, it appears that the tasks varied in objective and subjective complexity (Singer at al., 2013). Some words that we considered relatively simple to search for generated several queries, such as *galette des rois, el roscón de reyes* (Task 1) and *tarkkaavaisuustaidot* 'concentration capacity'(Task 2), whereas some complex words generated fewer queries than expected, such as *le CAPES, el copago* (Task 1) and *musisoiva* 'instrument-playing'(Task 2). The subjective complexity is thus also observed in the long search paths related to seemingly simple words when students tried to understand the functionality of the word in a specific context.

4.3.3 Information seekers' profiles

Regarding the individual paths taken, our study reveals different profiles of information seekers in both groups. We might refer to them as *go straight forward-seekers*, *advanced seekers*, and finally, *rely on your own intuition-seekers*—those who trust in their own language skills, knowledge, and judgment. Go straightforward-seekers just checked the dictionary and maybe one other source (see example 1 above). The paths of the advanced seekers were highly diverse, including both queries made in dictionaries and several pages on internet, to find the best alternative available. Examples 2 and 3 reflect advanced seekers' paths:

(2) *copago* (Test 1: economy): M (Spanish) + ML (RAE) + G (Spanish) + G (Spanish) + M (English) + M (Spanish) + O (Eur-Lex) + G (Finnish) + G (Spanish + Finnish) + O (ProZ.com) = 11 different sources used.

(3) *navaja* (Test 1: animals or biology): M (Spanish) + O (IATE) + ML (RAE) + M (Spanish) + M (Spanish) + M (Spanish) + G (image, Spanish) + G (image, Finnish) = 8 different sources used.

In both examples, the students did find the correct answer for their query. Examples reveal how complex the search paths might be. Furthermore, the search paths characterize clearly the mindset of an advanced seeker. Advanced seekers were not content with the first option they found, used different sources, and attempted to confirm the suitability of their answer. Finally, in Group B those who relied on their own intuition were usually students who had, for example, spent a longer time in a Spanish-speaking country, whereas in Group A, one's own expertise was often related to the previous knowledge of another language such as English. Example 4 shows the typical trail of an information seeker relying on one's own knowledge apparently without hesitation, whereas Example 5 illustrates the use of metaknowledge acquired from prior learned languages:

(4) *las migas* (Test 1: gastronomy): [My] Own [knowledge].

(5) *aivovaste* (Test 2): I remembered that the word corresponding to *aivovaste* is *neural response* in English, and so I decided to use Google and to see what results the key word *réponse neurale* would give.

The student in Example 4 was unsuccessful with the translation, whereas the student in Example 5 succeeded in finding the right answer.

4.3.4 Impact of pedagogical intervention

The results suggest that the instructional intervention had an effect on information seeking and its strategies, which can be seen in the diminished use of Wikipedia in Task 2 in both groups. The intervention might thus have had an effect on this tendency because the awareness-raising search tasks also included the evaluation of the trustworthiness of this open-source encyclopedia (Dobson & Willinsky, 2009; Rowley & Johnson, 2013). Increased use of the online dictionary MOT in both groups and other dictionaries in Group A most likely reflects the nature of the second task since it can be classified as a complex information-search task in which the students were asked to translate a text into a foreign language. In fact, the results indicate that in both tasks, MOT was the first step taken in the process of searching adequate solutions for the words in Group B. In Group A, for its part, the tendency of the search path was slightly different. In both tasks, the participants started their search most often with Google or other dictionaries, even if the use of MOT also increased in this group.

5. Discussion: Strategies for Finding Information

As outlined previously, digital literacy consists of two dimensions in which participants use different strategies. The first dimension is the relevant use of search engines. This was analyzed in both studies A and B. The first overall strategy was that all the participants used Google, which was forced in Study A and indicates that they formulated a search phrase containing one or more words and typed it in the text box. Advanced search methods related to the formulations also included the use of the different search possibilities available in Google. Among these, the participants used Google Images (for the questions related to plants), Google Maps (for the question related to a geographical location) and Google Translator (Study B). For the group working in Finnish L1 (Study A) or translating from or to Finnish L1 (Study B), the participants also used the language preference option and searched for pages written in Finnish or located in Finland.

The second dimension of digital literacy is competencies that are directly language related or other semiotic competencies with digital texts. Here, the selection and finding relevant information in another language when using Wikipedia was a major strategy. Participants went to the Wikipedia article dealing with the topic in French and then changed the language within Wikipedia to open the corresponding article in Finnish, if it existed. If the article did not exist, the participants might access an article in another language they were familiar with (usually English).

Furthermore, another strategy was related to the use of a third, intermediate, language to find some information about the topic. Latin was used as an intermediate language especially in the two questions dealing with plants (Study A) and animals or biology (Study B). There, the participants would first find a page about the topic in the language of the question (Study A) or the source language (Study B), then look up the scientific (Latin) name of the plant or animal on that page, and finally use the Latin name with the language preference option to find the name (and its taxonomic family) in the target language. In Study A, they might also add, in their query, keywords such as "glossary" or "dictionary" (in French or in Finnish) to find a page that might have the term in both languages. This strategy was used especially in the question asking to indicate a medical term that designates a malformation of the jaw.

6. Conclusion

We can conclude from Studies A and B that some of the results corroborate earlier studies. Searching for information on Google and online search engines were the most widely used information seeking strategies, even when the use was not forced (Massey & Ehrensberger-Dow, 2011). However, the students developed several individual search paths to complete the tasks. Advanced search methods included the use of the different search possibilities available in Google (Google Images and the language preference option). On several occasions, students used multilingual search strategies, including, for example, English, to find the appropriate answer or translation equivalent or a third additional language such as Latin to search for words in specific domains such as plants and animals (see Al-Eroud et al., 2011). English was systematically used by some participants who were first-year students of French or had French as their minor subject. In this case, two reasons are suggested as possibilities: they felt more comfortable with English than with French and more information elements are posted in English on the web (see Aula & Kellar, 2009).

To answer the first two research questions, as outlined above, the seeking behaviors were different regarding the different tasks in both studies, and this was partly due to the language background of the participants or the language in which queries were made. In other words, in Study A, information seeking in one's first language was faster than in L2, and the participants who had to change language from French into Finnish (group FI-FI) formulated more queries and used more keywords than the groups working in French, whereas in Study B, the results indicated partly different results according to the Group A (French) and Group B (Spanish) in both tasks even if the general tendencies were quite similar. In Study A, task complexity, considered as an objective measure, had a significant effect on task completion time and the number of formulations. In Study B, however, complexity was revealed to be partly a subjective measure (see Singer et al., 2013).

To answer the third research question about the impact of the awareness raising, we can conclude that it has a positive effect on digital literacy learning process (see Macaro & Mutton, 2009; Woore, 2010). It appears that the awareness raising during the intervention in Study B made the students evaluate the use of sources, and this was revealed in the way the students explicitly evaluated their findings and the trustworthiness of the sources used in Task 2 in comparison with Task 1 (see Dobson & Willinsky, 2009; Rowley & Johnson, 2013). In addition, the impact of the intervention might be observed in our study in the decreased use of Wikipedia and in the increased use of monolingual dictionaries. Study B also revealed that in Group B (Spanish), unlike in Group A (French), the overall number of queries decreased in Task 2, but the Google queries slightly increased. It is conceivable that instead of referring to Wikipedia explicitly in Task 2, these students might have started with Wikipedia and then have consulted the sources or references mentioned at the end of a Wikipedia article which may have lead them to other Google sources. After all queries, the students might have mentioned only the more reliable sources in their opinion. All these findings allow us to suggest that this type of processing approach with a pedagogical intervention can be used as a pedagogical tool to raise awareness of digital literacy.

In regards to the limitations of the two studies, in Study A, changing language had an (unwanted) effect on task complexity. That is, some easy questions turned out to be difficult when the information had to be found in Finnish because information on certain subjects was not as widely available in both French and Finnish (for example, on where bog rosemary can be found *in France*). This was demonstrated by longer completion times. In Study B, the study design as a mainly small-scale qualitative study prevents any generalization to the wider population. The limited sample did not allow more sophisticated quantitative statistics than basic correlations. The results were affected by individual variation, which was partly due to the heterogeneity of the groups (different age groups and/or proficiency levels and different background factors related to the course). Further research is required to determine whether the pedagogical intervention and the strategies described here are characteristics of other similar learning contexts. It is possible that the participants' behavior was influenced by the time on task, especially in Task 1. The use of logging user behavior design, such as that in Study A, combined with a think-aloud method could make the use of explicit queries more reliable.

In this type of learning context, the expected learning outcomes include students who are capable of engaging themselves in different types of language activities in different sociocultural and virtual contexts and are capable of analyzing and interpreting these activities. On a general level, the learning outcomes include aptitude for critical and scientific thinking, as well as knowledge production. Although today's students belong to a generation sometimes viewed as digital natives, their internet skills still have to be honed in their future professional areas.

References

Al-Eroud, A. F., Al-Ramahi, M. A., Al-Kabi, M. N., Alsmadi, I. M., & Al-Shawakfa, E. M. (2011). Evaluating Google queries based on language preferences. *Journal of Information Science, 37*(3), 282–292. doi: 10.1177/0165551511403383

Aula, A., & Kellar, M. (2009). Multilingual search strategies. *CHI 2009*, 3865–3870. doi: 10.1145/1520340.1520585

Aula, A., Khan, R. M., & Guan, Z. (2010). How does search behavior change as search becomes more difficult? *CHI 2010*, 35–44. doi: 10.1145/1753326.1753333

Barton, D., & Lee, C. (2013). *Language online: Investigating digital texts and practices*. London and New York: Routledge.

Chevalier, A., & Tricot, A. (2008). *Ergonomie des documents électroniques* [Ergonomics of electronic documents]. Paris: Presses universitaires de France.

Cope, B., & Kalantzis, M. (2010). New media, new learning. In D. R. Cole & D. L. Pullen (Eds.), *Multiliteracies in motion* (pp. 87–104). New York and London: Routledge.

Dervin, F., Johansson, M., & Mutta, M. (2007). Écriture académique: collaboration multimodale et co-constructions identitaires en FLA [Academic writing: multimodal collaboration and identity co-constructions. *Synergies, 2*, 93–106.

Dobson, T., & Willinsky, J. (2009). Digital literacy. In D. R. Olson & N. Torrance (Eds.), *The Cambridge handbook of literacy* (pp. 286–312). New York: Cambridge University Press.

Dommes, A., Chevalier, A., & Lia, S. (2011). The role of cognitive flexibility and vocabulary abilities of younger and older users in searching for information on the Web. *Applied Cognitive Psychology, 25*(5), 717–726.

Gilster, P. (1997). *Digital literacy*. New York: John Wiley & Sons.

Guitert, M., & Romeu, T. (2009). A digital literacy proposal in online higher education: The UOC scenario. *eLearning Papers, 12*, 1–15.

Gutiérrez, A., & Tyner, K. (2012). Media education, media literacy and digital competence. *Comunicar, 3*(1)*8*, 31–39.

Jewitt, C. (2006). *Technology, literacy, learning. A multimodal approach.* London and New York: Routledge.

Johansson, M. (2014). *Reading digital news.* Manuscript submitted for publication.

Johansson, M., & Dervin, F. (2007). Choix en matière de curriculum à l'université finlandaise et expertise à l'oral [Choices in curriculum materials in the Finish university and oral proficiency]. *Travaux de didactique du FLE de l'Université Paul Valéry, 57*, 27–46.

Kim, J. (2009). Describing and predicting information-seeking behavior on the Web. *Journal of the American Society for Information Science and Technology, 60*(4): 679–693. doi: 10.1002/asi.21035

Kress, G. (2003). *Literacy in the new media age.* London: Routledge.

Macaro, E., & Mutton, T. (2009). Developing reading achievement in primary learners of French: Inferencing strategies versus exposure to "graded readers". *Language Learning Journal, 37*(2), 165–182. doi: 10.1080/09571730902928045

Mackey, T. P., & Ho, J. (2005). Implementing a convergent model for information literacy: Combining research and web literacy. *Journal of Information Science, 31*(6), 541–555. doi: 10.1177/0165551505057018

Mangenot, F. (2012). Ecrire avec l'ordinateur: du traitement de texte au web social [Writing with the computer: from word processing to the social web]. *Le français dans le monde, 51*, 107–116.

Marcoccia, M. (2012). Conversationalisation et contextualisation: deux phénomènes pour décrire l'écriture numérique [Conversatinalization and contextualization: two phenomena for describing digital writing]. *Le français dans le monde, 51*, 92-106.

Martin, A. (2005). DigEuLit—A European framework for digital literacy: A progress report. *Journal of eLiteracy, 2*, 130–136. Retrieved from http://www.jelit.org/

Massey, G., & Ehrenseberger-Dow, M. (2011). Investigating information literacy: A growing priority in translation studies. *Across Languages and Cultures, 12*, 193–211.

New London Group. (1996). A pedagogy of multiliteracies: Designing social futures. *Harvard Educational Review, 66*(1), 60–92.

Newrly, P., & Veugelers, M. (2009). How to strengthen digital literacy. Practical example of a European initiative "SPreaD". *eLearning Papers, 12*, 1–9.

Olson, D. R., & Torrance, N. (Eds.). (2009). *The Cambridge handbook of literacy.* New York: Cambridge University Press.

Pinto, M. (2010). Design of the IL-HUMASS survey on information literacy in higher education: A self-assessment approach. *Journal of Information Science, 36*(1), 86–103. doi: 10.1177/0165551509351198

Pinto, M., & Sales, D. (2007). A research case study for user-centred information literacy instruction: information behaviour of translation trainees. *Journal of Information Science, 33*(5), 531–550. doi: 10.1177/0165551506076404

Pinto, M., & Sales, D. (2010). Insights into translation students' information literacy using the IL-HUMASS survey. *Journal of Information Science, 36*(5), 618–630. doi: 10.1177/0165551510378811

Rowley, J., & Johnson, F. (2013). Understanding trust formation in digital information sources: The case of Wikipedia. *Journal of Information Science, 39*(4), 494–508. doi : 10.1177/0165551513477820

Salmi, L., & Chevalier, A. (2010). *Recherche d'informations sur Internet – stratégies des internautes en fonction de leur profil* [Information research on the internet—internet users' strategies as a function of their profile]. Paper presented at the Conference of the European Society for Translation Studies (Tracks and Treks in Translation Studies), Leuven, Belgium.

Sharit, J., Hernandez, M. A., Czaja, S. J., & Pirolli, P. L. (2008). Investigating the roles of knowledge and cognitive abilities in older adult information seeking on the Web. *ACM Transactions on Computer-Human Interaction, 15*(1): article 3.

Singer, G., Norbisrath, U., & Lewandowski, D. (2013). Ordinary search engine users carrying out complex search tasks. *Journal of Information Science, 39*(3), 346–358. doi: 10.1177/0165551512466974

Thurlow, C., & Mroczek, K. (2011). Introduction: Fresh perspectives on new media sociolinguistics. In C. Thurlow & K. Mroczek (Eds.), *Digital discourse: Language in the new media* (pp. ix–xliv). New York: Oxford University Press.

UNESCO. (2008). *Teacher training curricula for media and information literacy*. Report of the International Expert Group Meeting. Retrieved from http://portal.unesco.org/ci/en/ev.php-URL_ID=27508&URL_DO=DO_TOPIC&URL_SECTION=201.html

Woore, R. (2010). Thinking aloud about L2 decoding: An exploration of the strategies used by beginner learners when pronouncing unfamiliar French words. *Language Learning Journal 38*(1), 3–17. doi: 10.1080/09571730903545210

Yaari, E., Baruchson-Arbib, S., & Bar-Ilan, J. (2011). Information quality assessment of community generated content: A user study of Wikipedia. *Journal of Information Science, 37*(5), 487–498. doi: 10.1177/0165551511416065

Chapter 11

New Literacy Practices and Plagiarism: A Study of Strategies for Digital Scrapbooking

MARTINE PETERS
Université du Québec en Outaouais (Canada)

MARY FRANKOFF
Cegep Heritage College (Canada)

Abstract

> Students nowadays often use various types of information, text, video, images, and audio found on the web in their writing assignments. They use a variety of digital-scrapbooking strategies to do so: finding information, evaluating it, and analyzing it in order to determine its relevance to their topic, taking notes, and then integrating the information through paraphrasing or quotations in their own work. Some will reference the authors whose work they have used; others will plagiarize, omitting any mention of their sources.
>
> In this survey-driven study, third-year students registered in a business English program at a Chinese University were surveyed about their information-searching strategies and their digital-scrapbooking strategies for their writing assignments. Likewise, students' perceptions of what they were taught about digital-scrapbooking strategies and copyright were examined. Results show that students' writing habits have changed; even though they frequently use the web to find information for their assignments, they are not as efficient as they could be in their use of digital-scrapbooking strategies. While their teachers might tend to neglect teaching digital-scrapbooking strategies, such as paraphrasing, they do teach referencing, and they seem to warn students about plagiarism.

1. Introduction

Before the turn of the century, being literate meant knowing how to read and write (Weis, 2004). Today, being literate implies a whole new set of skills: being able to access, analyze, and evaluate information in order to produce messages across a variety of contexts, including the internet. Thus, the ability to communicate efficiently through writing is an essential skill in order to live and flourish in the 21st century.

In the school system, the introduction of technology has changed how students learn and write. Technology has become a part of students' academic toolkit (Gibbons, 2007) as they have adopted various ways to communicate in writing, from the more traditional word processor to email, blogs, forums, and text messages. Learners have access to a wide variety of information—text, video, images, and audio on the internet that they can use as source material for their writing tasks. Students use digital-scrapbooking strategies to collage this information to write their assignments. They find websites related to the desired topic and analyze and evaluate them in order to select pertinent information. They then take notes and assimilate the various bits and pieces of information in order to integrate it into their own writing. While most teachers consider using information in this way to be plagiarism, there are others who consider that writing assignments in this manner demonstrates new and innovative ways to use the resources available to them (Ryberg & Dirckinck-Holmfeld, 2008). Information found on the web can be integrated in a student's work as long as it is quoted or paraphrased and the author is referenced; otherwise, we consider it to be plagiarism.

In this chapter, we examine how college students find and use information from the web for their writing assignments. We also investigate both the origins of students' digital-scrapbooking strategies and their knowledge about plagiarism. In the section below, we define the four key concepts for this paper: digital literacy, information seeking, digital-scrapbooking strategies, and plagiarism. We then explain the method for the study and present our results. Finally, we conclude with recommendations for teachers.

2. Background

2.1 Digital Literacy

At the beginning of the 20th century, the concept of literacy was associated with the ability to write your name. Later, the definition was broadened to include reading. In 1970, the concept of information literacy appeared, and it included the ability to find, evaluate, and use information (Weis, 2004). Then towards the end of 1990, the concepts of digital literacy, technological literacy, or online literacy surfaced (Dhillon, 2007). According to Jenkins, Purushotma, Clinton, Weigel, and Robinson (2006), the traditional skills of reading and writing constitute the foundation of the new skills of digital literacy that rely on various types of documents. Digital literacy skills today are not limited to written texts but also include images, sounds, animations, and videos. Students must become competent not only to read multimedia texts but also to navigate in a confusing and complex cyberspace in order to find what they seek (Brown, 2000).

Jenkins et al. (2006) argue that digital literacy skills include social skills, that is to say, they are more than a mere collection of skills used for the acquisition of information or for personal expression. This implies changes in working habits, values, and practices that are now more participatory and collaborative. Students need to develop new ways to interact with the sources of information available through the use of technology (Lankshear & Knobel, 2007). According to these authors, new textual forms are constantly emerging, and they are less often published on paper, less often written by a single author, and less individual. The need for expertise in order to be published is also disappearing. Indeed, since most writers can now be read by the general public simply by displaying their text on the internet, there is no longer a need to be an expert or a writer. Lessig (2004) stated that currently, information flows much more freely and is therefore free to be used and reused. "Writing is becoming ever more central and crucial to the world of work, with computers on every desk, email and the Internet adding to the world's written work in almost epidemic proportions" (Davies & Birbili, 2000, p. 430). This abundance of information does not do away with the obligation to quote sources. With more and more people writing, the study of digital literacy skills is an emerging area of inquiry. Researchers are interested in how texts are being written, and specifically in our case, how information is found and used in the texts being produced by students.

2.2 Information-Seeking Strategies

There is so much information available on the internet that today's students automatically tend to search the web when they are looking for information (Foster & Gibbons, 2007). The ability to be instantly linked to the internet at all times is expected. Statistics show that 71% of Canadian students use the internet for their schoolwork (Clark, 2001). According to Burton and Chadwick (2000), 66% of students questioned reported never having been trained on how to evaluate information found on the internet. Many students have expressed the wish to be trained by their teachers to evaluate the relevance, value, and credibility of information they find on the internet (Roy, 2009). Manual (2002) found that students who search for information on the internet have typically scan rather than read information. They skip from one website to another, and many teachers worry about the small amount of time used by students to search for and read information that will be the most relevant for their written assignments (Loertscher, 2008).

Foster and Gibbons (2007) have found that students judge the validity of the site by its popularity, while Fitzgerald's research (2004) established that students attribute the same validity to sources found online that they do to printed resources. This can be a problem since numerous websites contain incorrect or misleading information. Burton and Chadwick (2000) mention that the first three criteria high school and college students use when seeking information to write a paper are all based on the ease with which information can be found, obtained, and understood. This leads to what Lessig (2004) has called the culture of "copy and paste," facilitated by the easy access to information with technology.

2.3 Digital-Scrapbooking Strategies

Today, once students have found the information needed for their writing assignments, they remix the information into a new text. Ryberg and Dirckinck-Holmfeld (2008, p. 145) have used the metaphor of *patchworking* to describe the way students will assemble different pieces of material to create a new text. However, patchworking refers to the use of only one type of material. We therefore prefer the term *digital scrapbooking* since it accounts for the various types of information students will use (e.g., text, audio, video, and images) to write their paper. According to Ryberg and Dirckinck-Holmfeld (2008), students will combine, assemble, modify, and weave this information into a whole new creation. Moreover, the authors explain that these creative processes require critical thinking on the part of learners.

For some students, the process of digital scrapbooking may seem quite natural, and it involves much more than just copying and pasting information. It is a constructive, creative, and productive process since the text is constantly being amended by adding new information. Carrington and Marsh (2008) argue that increased access to information in various forms will allow for greater student creativity. Lankshear and Knobel (2006) argue that technology facilitates this new way of writing, and students' ability to remix, assemble, reformulate, and recontextualize therefore becomes very important.

However, this new way of learning and expressing oneself must be accompanied by the development of skills to analyze and evaluate the relevance of sources to enable individuals to make effective choices regarding the information they need. Developing the ability to make sound judgments is essential to being a good "scrapbooker" (Brown, 2000). According to Jenkins et al. (2006), there is another digital-scrapbooking strategy: appropriation. According to these authors, students must undo what the original authors of the text, video, and images have created by extracting what they understand and what they want to use into their own creation.

For many teachers, students who use digital-scrapbooking strategies to write papers choose the easy way, one which does not require critical judgment and comes dangerously close to plagiarism. Ryberg and Dirckinck-Holmfeld (2008), however, documented evidence to show that, when using digital-scrapbooking strategies, students have opportunities to use complex processes—including critical thinking skills—to produce a catchy new text.

According to Carrington and Marsh (2008), teachers can no longer limit themselves to the teaching of writing, but must also instil in their students a set of skills leading to the creation of new content by using various information collected on the internet. Olher (2009) explains that "[g]eneral literacy means being able to read and write the media forms of the day, which currently means being able to construct an articulate, meaningful, navigable media collage" (p. 10). Olher also stresses the importance of integrating traditional literacies and emerging literacies, which rely heavily on the technique of "remix culture" (see also Lamb, 2007).

The challenge for teachers lies in the overlap of two cultures, that of the "remix culture" and that of the school system which prohibits plagiarism. The perceptions of students and teachers with regard to this new way of creating texts differ greatly. Students who use technology and digital-scrapbooking strategies to complete their school assignments see this procedure as a natural practice (Jukes, 2008). Teachers, on the other hand, believe this is not a new creation, but instead plagiarism. Furthermore, they believe it violates copyright laws as well as the principles promoted by the school (Giezendanner, 2007). However, several authors (Lankshear & Knobel, 2007; Lessig, 2004) argue that "remix" or digital scrapbooking is creative since the choice of information and its arrangement make it into a new creation. These same authors are very explicit about the importance of teaching students how to cite their sources. Like many authors working in this field of specialization, Jenkins et al. (2006) warn against the dangers of failing to inform young people about the legal issues that this practice entails. Palfrey, Gasser, Simun, and Barnes (2009) reported that learning the importance of understanding and respecting the rights of authors should be done as soon as possible in school. Doing so will enable students to develop sensitivity and awareness in the exercise of their legal rights when they use other people's content in their digital scrapbooking. The authors explain that by encouraging creativity among students, they will develop a respect for other people's creations and will be more aware of the importance of assigning copyright to those whose works they have used in their own creations (Palfrey et al.).

2.4 Plagiarism

Definitions of plagiarism are plentiful, and Bretag and Mahmud (2009, p. 50) explain that the abundance of definitions is due to the "complexity of factors influencing its occurrence" including "the fact that plagiarism is often conflated with cheating and academic misconduct generally." For the purpose of this paper, plagiarism is defined as the appropriation of someone else's words or ideas and presenting that material as one's own. There are numerous types of plagiarism the most common ones cited in the literature are direct plagiarism, which is a simple copy and paste of exact words (Klausman, 1999; Walker, 2010); idea plagiarism in which students use someone else's work without quoting it (Stamatatos, 2001); trying to paraphrase an author yet in reality effecting few changes (Allan, Callagher, Connors, Joyce, & Rees, 2005; Kakkonen & Mozgovoy, 2010); and submitting a complete paper written by someone else (Park, 2003; Shei, 2005).

Unfortunately, plagiarism has become a quagmire for many teachers and schools. Research by Owunwanne, Rustagi, and Dada (2010) shows that plagiarism is considered acceptable and practiced widely by students (see also Gullifer & Tyson, 2010). Not surprisingly, research has also found that copyright issues are seldom taught in high school or college (McGowan & Lightbody, 2008). As a result, many students do not know how to quote or paraphrase properly, leading to unintentional plagiarism (Mittermeyer & Quirion, 2003). The use of the web as a source of information has facilitated plagiarism because students often have difficulty knowing what, when, and how to quote (Baruchson-Arbib & Yaari, 2004). Plagiarism is usually punished when detected, but many researchers indicate that punishment is not a long-term solution and that teaching about plagiarism and how to avoid it would be a better solution (Scanlon, 2003). Hollandsworth, Dowdy, and Donovan (2011) explain that students must develop the knowledge and skills to become good digital citizens, citizens who use technology appropriately and are aware of the consequences of their actions.

3. Research Questions

The present study was designed to investigate the habits of students with regard to their use of digital-scrapbooking strategies, including information-searching strategies and what they have been taught about these strategies. Specifically, the following questions guided this study:

1. How do students search for information for their writing assignments?
2. How do students use digital-scrapbooking strategies while writing assignments?
3. What are students taught about digital-scrapbooking strategies and copyright?

4. Method

Data were collected over the course of a term (i.e., 15 weeks) in a writing class. The participants were third-year students ($N = 90$), registered in a business English program in a Chinese university. Almost all students (91%) were between the ages of 20 to 25, and 93% of the students were women. More than two thirds (84%) had access to a computer at home or in the dormitory, and 81% indicated that they used their computer daily. When questioned about their teachers' use of instructional technologies in their classes, 63% stated that technology was used daily in class.

Three instruments were used to collect data from the participants. The first was a demographic questionnaire with five items about age, sex, first and second languages spoken, program of study, and number of years at the university level. The second questionnaire contained 11 items about the students' use of technology and their level of comfort and skills using technology. In the last questionnaire, participants answered 46 questions about their digital-scrapbooking strategies, their information-searching habits, their writing habits, and what they knew about plagiarism. The data from the questionnaire were analyzed with descriptive statistics.

5. Results

Results are organised in three sections below. The first examines how students conduct their information searches. The second concentrate on the digital-scrapbooking strategies used by students when writing their assignments, specifically planning their information search and reevaluating the information they found. The third section presents results pertaining to the help, information, or training with the digital-scrapbooking strategies students receive from their teachers on writing assignments that require information seeking.

5.1 Information Searches by Students

The first research question focuses on the frequency with which the students search for information during a term (see Table 1).

Table 1
How often do you gather information for your assignments?

Frequency	N	%
[Missing]	3	2.5
Never	0	0.0
1-2 times per term	9	7.5
3-5 times per term	29	24.2
More than 5 times per term	77	64.2
Always	1	0.8
At least once a week	1	0.8
Total	120	100.0

Over 64% of students indicated that they have to search for information more than five times a term (see Després-Lonnet & Courtecuisse, 2006; Le Douarin & Delaunay-Téterel, 2011). Thirty-two percent of the students search for information less than 5 times per term, which might be due to certain course content or to teachers not making information-seeking assignments. However, this represents only a third of the students; most students is search very frequently for information.

Slightly more than half the students indicated that they use technology daily to search for information (see Table 2).

Table 2
How often do you apply your technological skills for information seeking?

Frequency	N	%
[Missing]	1	0.8
Daily	61	50.8
3-4 times a week	44	36.7
5-10 times a month	12	10.0
Less than once a month	1	0.8
Never	1	0.8
Total	120	100.0

Only two students indicated rarely using or not using technology to search for information. Students seem to look online for information because it is easier to find and more easily accessible than in books (Agosto, 2002; Le Douarin & Delaunay-Téterel, 2011; Thompson, Schmidt, & Davis, 2003). The ease and the accessibility of information on the web might be why all of the students expressed having positive or very positive personal feelings when questioned about information seeking on the internet (see Table 3).

Table 3
What are your personal feelings towards information seeking on the internet?

Frequency	N	%
[Missing]	2	1.7
Very positive	85	70.8
Positive	33	27.5
Negative	0	0.0
Very negative	0	0.0
Total	120	100.0

This "Google generation" (Le Douarin & Delaunay-Téterel, 2011) is clearly comfortable using technology and the web to search for information. Once students have found the information they need, how do they integrate it into their writing assignments?

5.2 Digital-Scrapbooking Strategies Used While Doing Writing Assignments

In this section, we examine two digital-scrapbooking strategies students use when writing assignments: planning their information search and reevaluating the information found.

Writing outlines or drafts is not a priority for almost a quarter of the students (23%) questioned, and only a third do so once or twice a term. However, 37% of the students write a draft or an outline three times or more during a term (see Table 4).

Table 4
When completing an assignment that requires searching for information, do you prepare a draft or outline to guide you?

Frequency	N	%
[Missing]	3	2.5
Never	28	23.3
1-2 times per term	40	33.3
3-5 times per term	19	15.8
More than 5 times per term	27	22.5
Sometimes, very irregular	2	1.7
Always	1	0.8
Total	120	100.0

When preparing an assignment that required information seeking, students might choose not to prepare an outline because they know what information they need to write their paper and do not feel the need to prepare to search. Agosto

(2002) indicates that many students always use keywords when searching for information and so have only one method when searching for information. It is possible then that they do not feel the need to prepare for their search with an outline since their method never varies (Boubée, 2011). Those who do write outlines might do so because they use a variety of searching methods such as advance searches (Kennedy & Judd, 2011) and wish to be better prepared for their information search.

Once they have found information and have starting writing their assignments, do students reevaluate the information to validate its pertinence? Almost all of the participants (94%) indicated that they do reevaluate the information gathered at least once per term (see Table 5).

Table 5
How often do you reevaluate the information you have gathered for your assignments?

Frequency	N	%
[Missing]	3	2.5
Never	7	5.8
1-2 times per term	37	30.8
3-5 times per term	43	35.8
More than 5 times per term	29	24.2
Always	1	0.8
Total	120	100.0

Students might feel the need to reevaluate the information because they did not find the correct information or information that otherwise seems valid (Agosto, 2002; Le Douarin & Delaunay-Téterel, 2011). There is so much information on the web, and the technical skills they have acquired outside of formal learning contexts are sometimes difficult to transfer in a school context (Fluckiger & Bruillard, 2008) in order to properly evaluate their sources of information (Le Douarin & Delaunay-Téterel, 2011). If certain skills acquired outside of school are not easily transferable to a school setting, what of the skills acquired in school?

5.3 Help, Information, or Training with the Digital Scrapbooking-Strategies Students Receive from Teachers

Students might need extra help, information, or training from their teachers when writing assignments. How often do they receive it? The results displayed in Table 6.

Table 6
How often do you seek help from your instructor to complete an assignment that requires searching for information?

Frequency	N	%
Missing	3	2.5
Never	16	13.3
1-2 times per term	49	40.8
3-5 times per term	27	22.5
More than 5 times per term	24	20.0
If instructions are unclear	1	0.8
Total	120	100.0

Approximately 53% of the students rarely (once or twice a term) or never seek information from their teacher. This might be because they feel confortable searching for information or simply because they have understood the assignment. Yet almost 43% of the students ask for help, perhaps because they need to be reassured that they are heading in the right direction or lack self-confidence (Eccles & Wigfield, 2002). Another possible interpretation is that the subject matter is difficult for them. Others might need help with technology to complete the assignment (Tabary-Bolka, 2009) if there is little information available about the topic.

When asked if they had been taught how to complete a writing assignment, 17% of the participants chose *never*, while another 50% selected *1-2 times per term* (see Table 7). Only 25% of the participants stated that teachers give instructions sometimes (*3-5 times per term*) or often (*more than 5 times per term*).

Table 7
Do your teachers show you how to write a research paper?

Frequency	N	%
[Missing]	4	3.3
Never	20	16.7
1-2 times per term	60	50.0
3-5 times per term	18	15.0
More than 5 times per term	12	10.0
Outline	1	0.8
Once	2	1.7
Taught in 1st and 2nd year	2	1.7
I don't write any	1	0.8
Total	120	100.0

It seems that these students' teachers give some training on how to write a research paper but not very regularly; only 25% of the students mention that this is done more than 3-5 times a term. It is quite possible that teachers expect students to have learned this prior to college since the high school program has cross-disciplinary competencies specific to writing research papers (Ministère de l'Éducation, 2006). However, since those competencies are generic and not subject specific, it is possible that it is not addressed in detail in high school. College students, if they are not taught how to do research, then have to "figure it out" or ask for help when they need it. Unfortunately, our results show that students rarely ask for help.

Table 8 shows the same pattern when students are asked if their teachers show them how to evaluate the information they find.

Table 8
Do your teachers show you how to evaluate information you find?

Frequency	N	%
[Missing]	3	2.5
Never	42	35.0
1-2 times per term	44	36.7
3-5 times per term	20	16.7
More than 5 times per term	8	6.7
Methods	1	0.8
Yes	1	0.8
Once	1	0.8
Total	120	100.0

The number of students (42; 35%) who say they have never received some form of training to evaluate the information they find is much higher than it should be in an ideal setting. This means that approximately one third of the students have to learn how to evaluation information by themselves, and many do so; 94% of the participants indicated that they reevaluate at least once a semester (see Table 5 above). Another one third of the students (37%) mention having been shown how to evaluate the information they find once or twice a semester, which echoes Hayes, Hurtt, and Bee's (2006) recommendation that all teachers should give training at the beginning of the year on information evaluation.

The proportion of students who reported never having been taught to paraphrase is slightly lower (33%) than those who indicated they have never been trained to evaluate information (see Table 9).

Table 9
Do your teachers show you how to paraphrase information you have found?

Frequency	N	%
[Missing]	3	2.5
Never	39	32.5
1-2 times per term	57	47.5
3-5 times per term	8	6.7
More than 5 times per term	9	7.5
Methods	1	0.8
Once	2	1.7
Taught in 1st and 2nd year	1	0.8
Total	120	100.0

Almost 50% of the students state that they have been trained to paraphrase once or twice a semester. However, only 15% of students declare that their teachers have shown them how to paraphrase more than 3 times a term. One possible reason is that college teachers expect students to have been taught how to paraphrase in high school and think that students only need a reminder, not a complete lesson, on paraphrasing. Unfortunately, Tabary-Bolka's results (2009) show that students do not know how to respect copyright and use authors' ideas without proper referencing. It is then surprising to see in Table 10 that teachers have shown the vast majority of participants (84%) how to give their references in their writing assignments at least once or twice per term.

Table 10
Do your teachers provide instructions to you on how to give references?

Frequency	N	%
[Missing]	2	1.7
Never	15	12.5
1-2 times per term	57	47.5
3-5 times per term	28	23.3
More than 5 times per term	16	13.3
When asked	1	0.8
Specify MLA or APA	1	0.8
Total	120	100.0

If teachers do not seem very inclined to teach information evaluation and paraphrasing, this cannot be said regarding explanations about plagiarism (see Table 11).

Table 11
Do your teachers explain what plagiarism is to you?

Frequency	N	%
[Missing]	2	1.7
Never	4	3.3
1-2 times per term	44	36.7
3-5 times per term	23	19.2
More than 5 times per term	42	35.0
Once	2	1.7
In the first term	2	1.7
Always	1	0.8
Total	120	100.0

It seems that most teachers frequently talk about plagiarism since only 3% of students claimed they had never heard about it from teachers. Thirty-five percent of the participants said that they heard about plagiarism more than five times per term. One student even stated — in the area provided for optional comments — that teachers always explain plagiarism.

The number of writing assignments students have to write every term can probably explain the high frequency of explanations given about plagiarism. Hayes et al. (2006) state that teachers usually talk about plagiarism on the first day of the term and subsequently every time students are given explanations about writing assignments. Teachers are aware of students' information search habits and how easy it is for them to copy and paste information from the internet. It seems to follow that warning students about plagiarism and explaining it may indeed help to prevent it. Teachers also deem it necessary to warn students of the consequences of plagiarism as shown in Table 12.

Table 12
Do your teachers explain what the consequences of plagiarism are to you?

Frequency	N	%
[Missing]	2	1.7
Never	1	0.8
1-2 times per term	53	44.2
3-5 times per term	23	19.2
More than 5 times per term	41	34.2
Total	120	100.0

According to the participants, teachers put more emphasis on warning them about plagiarism than on explaining what its consequences can be. The frequency is not as high when it comes to explaining about consequences, probably because teachers might not be aware of how to detect plagiarism (Pickard, 2006) or even how to apply the school policy when faced with plagiarism (Carpenter, Harding, Finelli, Montgomery, & Passow, 2006; Macdonald & Carroll, 2006)2006.

6. Discussion

The results presented here show that students frequently search for information to write their school assignments. That they are using technology to do so is hardly surprising. A study by Després-Lonnet and Courtecuisse (2006) has shown that students do not know how to find books at the library. Some will even go to Amazon or Wikipedia to find information about books and then head to the library to try and find them.

Students in our study indicated that they like using the internet to look for information. One possible reason is that they find it much more quickly online than in the library, especially in difficult circumstances.

> Classrooms are information-poor environments. Thirty copies of the same, outdated book is not good enough, and the yearly $200 available to the school's library for all the subjects for new acquisitions is not going to do it either. And, in these times of cutbacks, we can't expect the public library to be particularly responsive. Where else are kids going to get the information resources they need but from the Web? (Soloway & Wallace, 1997, p.11)

The internet includes websites that offer frequently asked questions (FAQs) and tips to help students find information. While there are thousands of websites available at all times of day and night, the school library can be quite limited (Després-Lonnet & Courtecuisse, 2006; Le Douarin & Delaunay-Téterel, 2011).

Some authors (Agosto, 2002; Bennett, 2012; Le Douarin& Delaunay-Téterel, 2011) claim that students have difficulties finding what they are searching for. This might explain why our participants frequently look for information. They might find information and realize, as they are writing their paper, that the information is not relevant, which might explain why they often feel the need to reevaluate the information they have found.

Another reason they might often look for information is the fact that most of them do not seem to plan what they are going to write, and so their search for information might not be as efficient as it could be. Making an outline or any kind of rough draft can help define the broad lines of a paper and subsequently orient the search for information. Unfortunately, this is not a scrapbooking strategy that is often used by our participants.

It would seem that the participants described in this chapter have developed some level of digital literacy skills. While few plan their writing assignments, they can still search for information, find what they need, and reevaluate this information as they write. However, students might not be as digitally literate as they could be. College teachers may expect their students to have learned how to search the internet for information and how to evaluate it in high school. This is possibly because some teachers may not actually have the necessary knowledge to teach the technical aspects of looking for information on the web (Fluckiger & Bruillard, 2008). Teachers must be comfortable with technology in order to teach students digital literacy and scrapbooking strategies. Our results show that students are much more frequently warned against plagiarising than they are helped with its prevention. Macdonald and Carroll (2006) frame this issue in the following way:

> [Students] have responsibilities as to what are inappropriate practices, but [in the UK's Quality Assurance Agency (QAA) code of practice, there is nothing] about the responsibilities of institutions to ensure that students have the necessary skills to avoid plagiarism, or on staff to consider how best to minimise the opportunities for it through their assessment practices. (p. 235)

Students need to be taught—not just once or twice a term, but as each time they prepare a writing assignment—about searching for information, paraphrasing, and referencing (Hayes et al., 2006). Teachers cannot limit themselves to teaching writing since they also need to instil in their students the skills needed to create new content with the information they find on the internet (Carrington & Marsh, 2008). Some teachers might too often assume that students have been taught either in high school or by other teachers; however, this, unfortunately, does not seem to be the case (Fluckiger & Bruillard, 2007).

In order for teachers to see the importance of teaching digital-scrapbooking strategies, they must be trained to do so. Pickard (2006) suggests that teachers need workshops to learn how to prevent plagiarism, and institutions must play a role not only in training teachers but also in putting forward clear regulations for teachers to follow (Carpenter et al., 2006; Macdonald & Carroll, 2006) in order to help teachers apply rules and teach students the digital skills they need to do a research paper, which would, in turn, diminish the level of plagiarism (McCabe & Treviño, 1993). According to Palfrey et al. (2009), students need to learn as early as possible how to respect copyright, and this implies teaching them digital literacy skills and scrapbooking strategies.

7. Conclusion

Further research in the area of digital-patchworking strategies needs to be done in order to understand why students are inclined to copy and paste rather than use these strategies to create their own writing assignments while giving credit to the authors whose content and/or ideas they use. Data from a variety of sources (e.g., interviews and writing journals) could certainly provide in-depth insights into specific new habits and practices that some students have developed.

These writing habits seem to have changed with the appearance of an abundance of information available on the web, and, as teachers, we need to be aware of these changes. Rather than lament the fact that many of our students are copying and pasting information in their writing assignments, we need to be proactive and tap into these new digital skills that students have acquired. It is our responsibility to foster the development of these skills by teaching digital-scrapbooking strategies in order to ensure that our students will become digitally literate and will be prepared for learning, working, and communicating in the 21st century.

References

Agosto, D. E. (2002). Bounded rationality and satisficing in young people's web-based decision making. *Journal of the American Society for Information Science and Technology, 53,* 16–27.

Allan, G., Callagher, L., Connors, M., Joyce, D., & Rees, M. (2005, April). *Some Australian persectives on academic integrity in the Internet Age.* Paper presented at the meeting of EDUCAUSE Australasia 2005, Auckland, New Zealand.

Baruchson-Arbib, S., & Yaari, E. (2004). Printed versus internet plagiarism: A study of students' perception. *International Journal of Information Ethics, 1,* 1–6.

Bennett, S. (2012). Digital natives. In Z. Yan (Ed.), *Encyclopedia of cyber behavior* (Vol. 1, pp. 212–219). Hershey, PA: IGI Globa.

Boubée, N. (2011, October). *Caractériser les pratiques informationnelles des jeunes: Les problèmes laissés ouverts par les deux conceptions « natifs » et « naïfs » numériques* [Characterizing the informational practices of young people : The problems left unresolved by two digital conceptions "native" and "native."] Paper presented at the meeting of Communications Rencontres Savoirs CDI, Rennes, France. Retrieved from http://www.cndp.fr/savoirscdi/fileadmin/fichiers_auteurs/Actes/Rennes_2011/NB-RencontresSavoirsCDI-oct2011.pdf

Bretag, T., & Mahmud, S. (2009). A model for determining student plagiarism: Electronic detection and academic judgment. *Journal of University Teaching and Learning Practice, 6,* 49–60.

Brown, J. S. (2000). Growing up digital: How the web changes work, education and the ways people learn. *Change, 32,* 10–20.

Burton, V. T., & Chadwick, S. A. (2000). Investigating the practices of student researchers: Patterns of use and criteria for use of Internet and library sources. *Computers and Composition, 17,* 309–328.

Carpenter, D. D., Harding, T. S., Finelli, C. J., Montgomery, S. M., & Passow, H. J. (2006). Engineering students' perceptions of and attitudes towards cheating. *Journal of Engineering Education, 95,* 181–194.

Carrington, V., & Marsh, J. (2008). Forms of literacy. *Beyond Current Horizons.* Retrieved from http://www.beyondcurrenthorizons. org.uk/wp-content/uploads/ch3_final_ carringtonmarsh_formsofliteracy_20081218.pdf

Clark, W. (2001). *Kids and Teens on the Net.* Ottawa, Canada: Statistics Canada, Government of Canada.

Davies, C., & Birbili, M. (2000). What do people need to know about writing in order to write in their jobs? *British Journal of Education Studies, 48,* 429–445.

Després-Lonnet, M., & Courtecuisse, J.-F. (2006). Les étudiants et la documentation électronique [Students and electronic documentation]. *Bulletin des Bibliothèques de France, 51,* 33–41.

Dhillon, M. (2007). Online information seeking and higher education students. In M. K. Chelton & C. Cool (Eds.), *Youth information-seeking behaviors II: Context, theories, models and issues* (pp. 165–205). Lanham, MD: Scarecrow Press.

Eccles, J. S., & Wigfield, A. (2002). Motivational beliefs, values, and goals. *Annual Review of Psychology, 53,* 109–132.

Fitzgerald, M. A. (2004). Making the leap from high school to college. *Knowledge Quest, 32,* 19–24.

Fluckiger, C., & Bruillard, E. (2008). TIC : analyse de certains obstacles à la mobilisation des compétences issues des pratiques personnelles dans les activités scolaires [ITC : analysis of certain obstacles to the development of personal competencies in schoastic activities]. In *Proceedings of the Colloque international de l'ERTé: L'*éducation à la culture informationnelle. Retrieved from http://archivesic.ccsd.cnrs.fr/docs/ 00/ 34/31/28/PDF/2008-10_-_ Fluckiger-27-CICI2.pdf

Foster, N. F. & Gibbons, S. (2007). *Studying students: The undergraduate research project at the University of Rochester.* Chicago, IL: Association of College and Research Libraries.

Gibbons, S. (2007). *The academic library and the Net Gen student.* Chicago, IL: American Library Association.

Giezendanner, F.-D. (2007). *Le plagiat dans les systèmes éducatifs* [Plagiarism in educational systems]. Geneva: Département de l'instruction publique, République et Canton de Genève. Retrieved from http://wwwedu.ge.ch/ sem/documentation/documents/plagiat.pdf

Gullifer, J., & Tyson, G. A. (2010). Exploring university students' perceptions of plagiarism: A focus group study. *Studies in Higher Education, 35,* 463–481.

Hayes, D., Hurtt, K., & Bee, S. (2006). The war on fraud: Reducing cheating in the classroom. *Journal of College Teaching & Learning, 3,* 1–12.

Hollandsworth, R., Dowdy, L., & Donovan, J. (2011). Digital citizenship in K-12: It takes a village. *TechTrends, 55,* 37–47.

Jenkins, H., Purushotma, R., Clinton, K., Weigel, M., & Robinson, A. J. (2006). *Confronting the challenges of participatory culture: Media education for the 21st century.* Cambridge, MA: MIT Press. Retrieved from http://mi tpress.mit.edu/sites/default/files/titles/free_download/9780262513623_ Confronting_the_Challenges.pdf

Jukes, I. (2008). *Understanding digital kids: Teaching & learning in the new digital landscape.* Retrieved from http://www.educationthatworks.net/ uploads/ 7/8/3/0/7830610/ understanding_digital_kids.pdf

Kakkonen, T., & Mozgovoy, M. (2010). Hermetic and Web plagiarism detection systems for student essays: An evaluation of the state of the art. *Journal Educational Computing Research, 42,* 135–159.

Kennedy, G. E., & Judd, T. S. (2011). Beyond Google and the "satisficing" searching of digital natives. In M. Thomas (Ed.), *Deconstructing digital natives. Young people, technology, and the new literacies* (pp. 119–136). New York, NY: Routledge.

Klausman, J. (1999). Teaching about plagiarism in the age of the Internet. *Teaching English in the Two-Year College, 27,* 209–212.

Lamb, B. (2007). Dr. Mashup, or why educators should learn to stop worrying and love the remix. *EDUCAUSE Review, 42*(4), 12–24. Retrieved from http://net.educause.edu/ir/library/pdf/erm0740.pdf

Lankshear, C., & Knobel, M. (2006). *New literacies: Everyday practices and classroom learning* (2nd ed.). London, England: Open University Press.

Lankshear, C., & Knobel, M. (2007). Sampling "the new" in new literacies. In M. Knobel & C. Lankshear (Eds.), *A New Literacies Sampler* (pp. 1–25). New York, NY: Peter Lang.

Le Douarin, L., & Delaunay-Téterel, H. (2011). Le « net scolaire » à l'épreuve du temps « libre » des lycéens [The "scholastic net:" proof against the "free" time of high school students] . *Revue Française de Socio-Économie, 8,* 103–121.

Lessig, L. (2004). *Free culture: How big media uses technology and the law to lock down culture and control creativity.* New York, NY: Penguin Press.

Loertscher, D. (2008). What works with the Google generation? *Teacher Librarian, 35,* 42.

Macdonald, R., & Carroll, J. (2006). Plagiarism—a complex issue requiring a holistic institutional approach. *Assessment & Evaluation in Higher Education, 31,* 233–245.

Manual, K. (2002). Teaching information literacy to Generation Y. *Journal of Library Administration, 36,* 195–217.

McCabe, D. L., & Treviño, L. (1993). Academic dishonesty: Honor codes and other contextual influences. *Journal of Higher Education, 64,* 522–538.

McGowan, S., & Lightbody, M. (2008). Enhancing students' understanding of plagiarism within a discipline context. *Accounting Education, 17,* 273–290.

Ministère de l'Éducation. (2006). *Programme de formation de l'*école québécoise: Enseignement secondaire, premier cycle [Curriculum of Quebec schools: Secondary Education, first cycle]. Quebec City, Quebec, Canada: Gouvernement du Québec. Retrieved from http://www.meq.gouv.qc.ca/DGFJ/dp/programme_de_formation/secondaire/prformsec1ercycle.htm

Mittermeyer, D., & Quirion, D. (2003). Étude sur les connaissances en recherche documentaire des étudiants entrant au *1er cycle dans les universités québécoises* [Study of the knowledge of document research by students entering the first cycle in Quebec universities]. In Conférence des recteurs et des principaux des universités du Québec. Retrieved from http://www.crepuq.qc.ca/documents/bibl/formation/etude.pdf

Olher, J. (2009). Orchestrating the media collage. *Educational Leadership, 66,* 8–13.

Owunwanne, D., Rustagi, N., & Dada, R. (2010). Students' perceptions of cheating and plagiarism in higher institutions. *Journal of College Teaching and Learning, 7,* 59–68.

Palfrey, J., Gasser, U., Simun, M., & Barnes, R. F. (2009). Youth, creativity, and copyright in the Digital Age. *International Journal of Learning and Media, 1,* 79–97.

Park, C. (2003). In other (people's) words: Plagiarism by university students—literature and lessons. *Assessment & Evaluation in Higher Education, 28,* 471–488.

Pickard, J. (2006). Staff and student attitudes to plagiarism at University College Northampton. *Assessment & Evaluation in Higher Education, 31,* 215–232.

Roy, R. (2009). *Les utilisateurs 12-24 ans: utilisateurs extrêmes d'Internet et des TI* [Users 12-24 years of age : extreme Internet and TI users]. Retrieved from http://www.csrs.qc.ca/fileadmin/user_upload/Page_Accueil/Parents/PDF/TIC_Jeunes/CEFRIO_GenerationC_Internet_TI_oct2001.pdf

Ryberg, T., & Dirckinck-Holmfeld, L. (2008). Power users and patchworking: An analytical approach to critical studies of young people's learning with digital media. *Educational Media International, 45,* 143–156.

Scanlon, P. M. (2003). Student online plagiarism: How do we respond? *College Teaching, 51,* 161–165.

Shei, C. (2005). Plagiarism, Chinese learners and western convention. *Taiwan Journal of TESOL, 2,* 97–113.

Soloway, E., & Wallace, R. (1997). Does the Internet support student inquiry? Don't ask. *Communications of the ACM, 40*(5), 11–16.

Stamatatos, E. (2001). Plagiarism detection using stopword n-grams. *Journal of the American Society for Information Science and Technology, 62,* 2512–2527.

Tabary-Bolka, L. (2009). « Culture adolescente vs culture informationnelle » : L'adolescent acteur de la circulation de l'information sur internet [Adolescent culture versus informational culture : The teenage distributor of information on the internet]. *Les Cahiers du numérique, 5,* 85–97.

Thompson, A. D., Schmidt, D. A., & Davis, N. E. (2003). Technology collaboratives for simultaneous renewal in teacher education. *Educational Technology Research and Development, 51,* 73–89.

Walker, J. (2010). Measuring plagiarism: Researching what students do, not what they say they do. *Studies in Higher Education, 35,* 41–59.

Weis, J. (2004). Contemporary literacy skills. *Knowledge Quest, 32,* 12–15.

Chapter 12

Preparing Future Foreign Language Teachers: The Role of Digital Literacies

JANEL PETTES GUIKEMA
MANDY R. MENKE
Grand Valley State University (USA)

Abstract

Given the increasing need for developing digital literacies in K-12 students, it is necessary to better understand how future teachers conceptualize digital literacies. With this in mind, the authors of this study set out to explore how future foreign language (FL) teachers consider digital literacies in their current and future instructional practice. The reflections of teacher candidates on instructional technologies presented as part of a technology videoconference in a FL methods course comprised the data set for this study. These future teachers expressed an overall enthusiasm about using technology, highlighting its potential to positively impact vocabulary, grammar, and culture learning, but they did not extend the instructional purpose to general digital literacies development. These findings underscore the importance of explicitly addressing digital literacies in a cohesive way throughout the FL teacher education curriculum.

1. Introduction and Background

Today's preservice teachers are part of a generation considered to be digital natives (Prensky, 2001; see also Benini & Murray and Lotherington & Ronda, this volume) and are assumed to be digitally literate individuals, familiar with a variety of technologies, able to efficiently search for and find information, compare and evaluate a variety of sources, and classify sources according to specific characteristics (Livingstone, van Couvering, & Thumin, 2005). In a world where electronic media are becoming increasingly ubiquitous and influential (Buckingham, 2008), both in daily life and educational settings, there is a need to foster K-12 students' "ability to understand and use information in multiple formats from a wide variety of sources," both digital and nondigital (Gilster, as cited in Pool, 1997, p. 6). The purpose of this study is to explore how FL teacher candidates envision the role of digital literacies and technology integration in language learning and in their own instructional practices.

Digital literacies have received increasing attention in FL education literature in recent years. This notion is conceptualized as more than technical competencies and the ability to think critically about information. Martin (2005) defines digital literacies as

> the awareness, attitude and ability of individuals to appropriately use digital tools and facilities to identify, access, manage, integrate, evaluate, analyze and synthesize digital resources, construct new knowledge, create media expressions, and communicate with others, in the context of specific life situations, in order to enable constructive social action; and to reflect on this process. (p. 135)

Digital literacies, a thread in the Common Core State Standards Initiative,[1] are developed in K-12 education through a system of shared responsibility, similar to the responsibility for addressing literacy being "predicated on teachers of ELA [English Language Arts], history/social studies, science, and technical subjects" (National Governors Association Center for Best Practices, Council of Chief State School Officers, 2010). Content area teachers, Language Arts teachers, and media specialists all must equip students with the language and skills necessary to read, write, speak, and listen across content areas in multiple formats, both digital and nondigital. The pedagogical endeavor of attending to digital literacies in the classroom is more than just integrating technology into instruction, it is "mentoring" students and creating "rich learning systems" (Gee, 2007, p. 138) that provide students with ways of understanding the cultural forms with which they are interacting both in school and outside of school (Buckingham, 2008). While a number of studies have focused on how teachers address digital literacies, research on how teacher candidates address them in the classroom is scarce, with most studies focusing on either literacy or technology integration. In an attempt to address the topic of digital literacies in teacher preparation in a holistic manner, this literature review focuses on commonalities across these two fields of research in an attempt to bridge the gap between them. It should be noted that the literature review pulls from the literature on both teachers and teacher candidates and focuses primarily on studies from content areas other than FL due to the limited amount of research specific to teacher candidates or the FL context.

Many studies point to the tendency of teachers to teach the way they were taught (e.g., Russell, 1997; Schifter, 1997; Scholz, 1995). Cuban (1995), Graham (2008), and Ottesen (2006) explain that this tendency holds true for technology in that the digital experiences of teachers, both within and outside of academic settings, impact their use of and instruction with technology in their classroom. For example, Graham suggests the teachers most willing to bring playful or social digital worlds into their teaching are those who are able to draw on their own "funds of knowledge" (González, Moll, & Amanti, 2005, p. ix) about digital

---

[1] The Common Core State Standards define core mathematical and literacy skills and knowledge all U.S. students need to develop in order to graduate high school prepared for college coursework and/or to enter the workforce. At the time of publication, 45 states, the District of Columbia, and four territories had adopted the Standards (Achieve, 2013).

worlds, having themselves experienced a sense of community and engagement via technology. In contrast, teachers who have not experienced "fun and delight" in digital worlds, often because they learned about it through self-study or schooling, view digital worlds and texts only as a necessary tool to "get on in the world of work" (Graham, p. 13) as opposed to a possibility for instructional engagement. In other words, teachers who have not experienced collaborative digital communities are less likely to use technology as an instructional tool and instead view it as an object of instruction.

Similar findings have been reported with respect to addressing multiliteracies in the classroom, defined as "the many and varied ways that people read and write in their lives" (Purcell-Gates, 2002, p. 376), whether it be print or nonprint media. Nahachewsky (2007) found that experienced Language Arts teachers continued to grapple with implementing practices that meaningfully addressed these "new literacies" in part because they lacked a deep understanding of what literacy instruction in a digital age should look like. One teacher attributed this in part to having had learning experiences far different from those of her students:

> These kids are used to thinking in ways that we never were. We were taught in rote fashion, in demand fashion ... so while they are maybe not as methodical or as organized as we would have been, they can multi-task and see a variety of influences coming into a main concept. Their paradigms are different. Rather than seeing things linearly on paper they open up their computer and like a hyper text there is another window that opens up and inside that and so on ... . And that's the way their brain must be hardwired — mine is not hardwired that way. (p. 360)

Sheridan-Thomas (2007) illustrates the complexity of impacting teacher candidates' practices and bridging the [literacy] theory-practice divide. She found that as a result of a content area literacy class, teacher candidates had conceptual knowledge of the complex, multifaceted nature of literacy and the potential disconnect between students' school-based and out-of-school literacy practices; nonetheless, as these teacher candidates planned instruction, their attention to multiple literacies was minimal and often confounded with differentiating for learning style preferences. Moreover, when multiple literacies were incorporated into instructional activities, it was often happenstance and not the result of intentional planning.

As one way of impacting the future instructional practices of teacher candidates, the teacher preparation literature argues the need for teacher candidates to experience technology and/or multiple literacies integration as a learner (Bucci et al., 2004; Burnett, 2011; Merkley, Schmidt, & Allen, 2001; Nahachewsky, 2007; Ottesen, 2006), yet the impact of technology-based learning experiences on teacher practices is also unclear. Some studies have found that opportunities to experience integrated, meaningful technology learning as a learner result in greater and more frequent technology incorporation in teaching (Dutt-Doner, Allen, & Corcoran, 2006; Mayo, Kajs, & Tanguma, 2005); however, according to Ottesen (2006), others such as Delange and Skedsmo (2004) and Jensen (2003) question the connection between using technology as a learner and then using those technology tools in teaching. Ottesen notes that although the available technologies might be the same or similar in the university and K-12 classroom setting, technology integration "is talked about and practised differently when the purpose for student teachers is to learn teaching, and when the purpose is to teach" (p. 278).

Interestingly, recent studies find that, although they are considered to be digital natives, teacher candidates may be less technologically expert than their students (Robinson & Mackey, 2006) and/or lack confidence when it comes to using technology for instructional purposes (see Bosch & Cardinale, 1993; Davis, 1993; Fratianni, Decker, & Korver-Baum, 1990; Heinich, 1991; Nahachewsky, 2007; Robinson & Mackey, 2006; Topp, 1996; Topp, Mortensen, & Grandgenett, 1995). This feeling of ill-preparedness is not without reason because developing technology-integrated curricular lessons has been identified as one of the most difficult tasks for classroom teachers (Conte, 1997; Driskell, 1999). While for some teacher candidates technological skill may be a concern, the greatest uncertainty rests in how to use technology in the classroom. As Dutt-Doner et al. (2006) assert, "meaningful technology integration is not so much a technical endeavor as a pedagogical one" (p. 66).

Most teacher preparation institutions emphasize meaningful technology and literacy integration in several courses; nonetheless, teacher candidates' emerging professional identities also play a role in how they operationalize knowledge and competencies gained from coursework. As Burnett (2011) argues, if understandings and experiences enter into conflict with certain kinds of teaching identities, they may be deemed irrelevant. Most teacher candidates strive to establish an identity as an "accountable practitioner" (Ottesen, 2006, p. 281), able not only to manage their students but also to empower them to achieve learning outcomes (cf. Shotter, 1984; Ottesen, 2007). Incorporating innovative tools or strategies, such as technology or digital literacy practices, into existing or established practices produces tension. This tension is revealed in the comments and reflections of teacher candidates and centers on the efficiency of using technology, uncertainty as to whether students are learning what they need to learn, and losing control over the teaching and learning process (Burnett, 2011; Ottesen, 2006). Through the comments of teacher candidates and mentor teachers alike, it becomes evident that technology in some cases is seen as an "add-on," a nonessential part of content learning and/or teaching. Both Burnett and Ottesen argue that teacher educators must invest in more than just the development of competencies but also in teacher identities, for "the extent to which new teachers draw on their digital experiences may be linked to how they see themselves both as technology-users and as teachers and how these interact" (Burnett, p. 435). It follows then that teacher candidates' practices are not solely influenced by coursework; they are additionally influenced by previous experiences, as well as beliefs and professional identities (Burnett; Graham, 2008; Ottesen).

This review has until now focused on research from teacher candidates in content areas other than FL education; there is a striking paucity of research related to FL teacher candidates' digital literacy practices and their attention to them in instruction. However, "it is precisely because literacy is variable and intimately tied to the sociocultural practices of language use in a given society that it is of central importance in our teaching of language and culture" (Kern, 2000, p. 25). While the literature on digital literacies or multiliteracies and technology is growing in the field of FL education (see Kassen & Lavine, 2007; Reinhardt & Thorne, 2011; Thorne & Reinhardt, 2008; van Compernolle & Williams, 2011), FL methodology textbooks still do not explicitly address digital literacies. Arnold (2013) examines the treatment of computer assisted language learning (CALL) in 11 of the most used methodology texts and finds that most of the textbooks weave the discussion of CALL throughout the text, integrating it into other topics; she interprets this to be a signal that "CALL can no longer be viewed as an optional add-on" (p. 233), in contrast to the reality painted by Ottesen (2006) and Burnett (2011). Although Arnold does not consider directly whether or how digital literacies are explored, she does note that textbooks emphasize the "planned and purposeful use of technology" (Shrum & Glisan, 2010, p. 451) by connecting the discussion of CALL to developing learners' communication skills in all three modes and linking learners with texts from other cultural groups as a means of exploring other groups' cultural practices, products, and perspectives. Through these connections the potential for addressing digital literacies exists, yet the topic is not explicitly addressed, allowing the separation between literacy and technology to continue. Aside from this analysis of textbooks used in FL methodology courses, to date, there has been no formal analysis of FL teacher candidates' understanding of digital literacies, and their attention to them in their instruction has yet to be explored.

The present case study, focused on an instructional project carried out in a FL teaching methods course during three different semesters, is an initial step toward filling this gap and beginning to understand how FL teacher candidates conceptualize digital literacies and how to incorporate them into the FL classroom. It attempts to answer the question of whether FL teacher candidates are "tuned in" to developing digital literacies in their future students. The guiding questions for this study were:

- Do digital literacies play a role as teacher candidates consider future instructional practice? If so, what is their role?
- How do they position themselves with respect to technology integration?

2. Method

The findings reported in this chapter are based upon data collected over three semesters from 32 teacher candidates pursuing a major in French, German, Latin, or Spanish secondary education. The number of teacher candidates varied from semester to semester; the number of participants and the language they planned to teach are presented in Table 1. All participants were enrolled in a FL methods course and were completing their first teaching practicum. As such, they spent four to five hours each day in a FL classroom at a secondary school, where they observed and assisted the cooperating teacher, interacted and worked with individual and small groups of students, and taught a 3-week unit.

Table 1
Teacher Candidate Participants by Semester

Fall 2011 $N = 16$	Fall 2012 $N = 9$	Winter 2013 $N = 7$
1 French, 2 German, 1 Latin, 12 Spanish	9 Spanish	1 French, 6 Spanish

The methods course, taught by a faculty member with expertise in second language acquisition and teaching, encompasses a wide range of topics in alignment with the National Council for Accreditation of Teacher Education Standards specific to the field of FL (American Council on the Teaching of Foreign Languages, 2002). One of the final assignments for the course is to design a thematic unit that integrates multiple modes of communication and focuses on meaningful language use, based on the *Standards for Foreign Language Learning in the 21st Century* (National Standards in Foreign Language Education Project, 2006). Technology must be integrated into at least one of the lessons, and its use must be accompanied by a rationale statement.

As part of the methods seminar, teacher candidates participated in a "conference in the classroom," in which a panel of university researchers and practitioners in FL acquisition presented instructional activities and research via Skype or ooVoo. During the conference, teacher candidates experienced technology's "dual role" (Fulton, 1998) since it was both the content of instruction and the tool for instruction. The conference had two major goals:

- to instill in teacher candidates the dispositions and skills necessary for thinking about technology, as well as
- to familiarize teacher candidates with the format of professional conference presentations and to provide an opportunity for dialogue within a professional discourse community.

Prior to the conference, the teacher candidates read the chapter, "Using technology to contextualize and integrate language instruction," from *Teacher's Handbook: Contextualized Language Instruction* (Shrum & Glisan, 2010) and also discussed different instructional technologies that they had either experienced as a student or observed in their current role as a teacher candidate. During the conference, each of three panelists shared a technological tool implemented in language instruction, discussing both the logistics of classroom implementation and the impact on student learning. Each semester two of the presenters were at remote locations, and one was on site. Furthermore, presenters represented multiple roles in FL education: several were researchers on technology and FL learning, one was a university language resource center director, and one was a high school FL teacher. The tools presented during each conference are organized by semester in Table 2. As can be seen, the technology varied from immersive, high-tech environments to form-focused practice and also to technology that required creative production. All the tools allowed for FL learners to engage in a variety of digital literacy practices.

Table 2
Tools Presented During the "Conference in the Classroom" by Semester

Fall 2011	Fall 2012	Winter 2013
Place-based mobile gaming for language learning	Place-based mobile gaming for language learning	A collaborative, open-access digital learning environment
Web-based exploration of culturally appropriate use of language	Chatrooms	Edmodo, Quia
Graphic novels	Graphic novels	Graphic novels

The teacher candidates were required to listen to each presentation, explore the technology presented, and then participate in the follow-up discussion in which they had the opportunity to address questions to panelists. After the conference, participants completed a short online survey with open-ended questions (see survey questions in the Appendix to this chapter) designed to explore candidates' reactions to the technological tools presented in the conference. Survey questions were intentionally open ended and broad so as to allow teacher candidates the opportunity to comment on what they felt was most pertinent to their instruction. The authors then reviewed, organized, and categorized all student comments from the survey and the panel discussion according to general and specific themes that emerged during the review and organization of the data.

3. Findings and Discussion

It became evident that the ways in which FL teacher candidates conceptualize digital literacies align closely with that of teacher candidates in other content areas (as identified in the literature review). In this section, we explore teacher candidates' perceptions of technology's effectiveness, how digital literacies tie into their vision of FL teaching, and how they position themselves with respect to technology integration.

3.1 Technology, Digital Literacies, and Foreign Language Teaching

From their comments about various uses of technology in the FL context, it is clear that these teacher candidates are themselves engaging in digital literacy practices as part of this videoconference; they feel empowered to analyze and evaluate the instructional technologies and their applications, to disseminate information to other teachers, and to propose adaptations and improvements. The pursuit of meaningful technology integration is central to their engagement in these digital literacy practices. Teacher candidates' focus is on the learning associated with each tool; they are not seduced by bells and whistles, and "embrac[e technology] only if there are substantial benefits to the learners" (Shrum & Glisan, 2010, p. 451). One teacher candidate addresses this notion directly:

(1) I think it can be confusing when [technology] is best and when we need to let it go. I think this is a really important point because our professors really push us to use a lot of technology in everything we do, but sometimes that isn't practical. I also liked that [the presenter] said it is not always efficient, but we shouldn't give up. (Fall 2012)

The instructional purpose behind a particular technology's use is at the forefront of teacher candidates' comments. They highlight a variety of skills that can be addressed through technology integration, many of which are encompassed within digital literacies. The future teachers emphasize features of (content-general) good pedagogy, such as creativity, interaction, and extending learning beyond the classroom, as they reflect upon the potential impact of tools on student learning.

(2) [Graphic novels][2] will be great to promote student collaboration, something completely new and innovative to encourage speaking, awareness of grammar, and creativity! (Fall 2011)

(3) For advanced classes, I would truly like to use these as I think they bring a new style of learning to the classroom. I also enjoyed finding different ways to implement games, such as World of Warcraft, in order to let students use their interests to learn the language. I think that next semester I would like to create a project where students can create their own game using the language that will let them be creative and innovative in class. (Fall 2011)

(4) [Chatrooms are] a great and effective way for students to think in the second language outside of class. (Fall 2012)

Technology's ability to tap into students' creativity, which can at times be limited by a lack of language proficiency in the FL classroom, is regularly commented on; this, in addition to technology's potential for collaboration and communication, is considered to be advantageous by teacher candidates. In this way, the instructional technology presented conforms to their understanding of good pedagogy and additionally brings something extra, something "new and innovative," into the classroom, serving to engage students in language learning.

Another advantage of technology identified by these future teachers is the access it provides not only to target language resources but also to native speakers, who fulfill the role of communicative partner, both in presentational and interpersonal (synchronous or asynchronous) communication. One participant specifically notes the importance of providing students with an authentic audience beyond the teacher and their classmates when in the presentational mode.

(5) I like what [presenter] had to say about making the [graphic novel] project for the public. Tell the students it will be viewed online, YouTube, or have an in-class viewing party. I think it would motivate students to put more into the project. (Fall 2012)

---

[2] The authors have identified the tool referenced in participants' comments; however, in some cases, the link to a particular technology is unclear.

Offering students a real audience for their work is discussed in FL methodology texts (e.g., Omaggio Hadley, 2000; Shrum & Glisan, 2010), yet this can prove challenging when the work is in a language not spoken in the immediate community. This particular teacher candidate suspects that using technology as a platform to host and make public student work is likely to increase student engagement with the content, potentially resulting in higher quality products and greater language learning. While other future teachers in this study do not explicitly address the notion of audience, they do sense that connecting learners to "a multiplicity of perspectives, a cacophony of voices, different and differing worlds, cultures, and languages" (Sawhill, 2008, p. 15) is both meaningful and exciting.

Similarly, teacher candidates emphasize technology's potential to connect learners with native speakers and advance learners' FL skills. Research suggests that learners value opportunities to interact with native speakers in real-world, authentic settings such as those available through technology (Lee, 2004); moreover, interactions with native speakers may offer learners opportunities to acquire new vocabulary and practice grammatical structures (Fernández Dobao, 2012; Lee, 2004).

(6) [Videoconferencing] would be a great way to have students speak with a native Spanish speaker and see how easy it can be! (Fall 2011)

(7) I think [place-based mobile gaming] is a great way to ... learn about another culture.... They have to know how to respond to someone who is very old-fashioned and uptight to get information as well as someone who is very laid back. (Fall 2011)

In the comments above, teacher candidates identify the learning outcomes that can result from spontaneous communication with native speakers, including learner confidence, cultural awareness, and pragmatic competence.

Interpersonal communication skills are critical for using technology to successfully build and maintain personal connections. In fact, most university students, including teacher candidates, invest significant time and resources maintaining these connections (Sawhill, 2008). It is this social aspect of technology that emerges as a key factor in teachers' willingness to embrace instructional uses of technology (Graham, 2008); when a technology application does not allow for personal connections, teacher candidates are hesitant to use it.

(8) I really do enjoy this incorporation of quizzes and games on the computer, but I also find it impersonal at times. (Winter 2013; Edmodo, Quia)

(9) I really enjoyed [the videoconference]. Like I said, this is something I can definitely see myself implementing in my future classroom. It's more personal than email ... (Fall 2011)

Such statements suggest that teacher candidates not only recognize but also highly value the personal connections that technology facilitates, demonstrating an awareness of this facet of digital literacies. Nonetheless, while teacher candidates reference multiple aspects of digital literacies, such as creativity, communication, and collaboration, they do not identify the FL teacher's role in developing students' ability to "appropriately use digital tools" (Martin, 2005). Based on these data, the focus of FL teacher candidates is FL learning, defined as grammar, vocabulary, the four language skills, and the three modes of communication as seen in statement (2) above and the comments below.

(10) I think that [graphic novels] will be educational and fun for my future students. I would like to use this at the end of a grammar section or with a specific vocabulary list. I also like how students can record their voices so it helps with writing, reading, and listening at the same time. (Fall 2011)

(11) I'm a big believer of interpersonal communication and I think that this online chat with students is a good way for them to practice. I'll use this strategy as a way to practice certain grammatical structures. (Fall 2012)

(12) [Graphic novels] would be beneficial in creating short stories when you need to emphasize vocab, grammatical rules or even culture! (Fall 2012)

While comments such as these reveal an awareness of developing vocabulary and grammar to facilitate communication, they reflect a somewhat narrow focus of FL teaching. Their focus centers largely on the Communication goal area of the *Standards for Foreign Language Learning in the 21st Century* with some mention of the Cultures goal area; they do not reference the possibilities technology offers for addressing the other goal areas: Comparisons, Connections, and Communities (National Standards in Foreign Language Education Project, 2006).[3] It is precisely the access to additional bodies of knowledge and distinct viewpoints that would allow teachers to engage their students in building new understandings from multiple sources and participating in social action. Digital literacy practices have the potential to tie into all five goal areas, whose overall purpose lies in preparing students to successfully navigate pluralistic societies, both at home and abroad (National Standards in Foreign Language Education Project). Involvement with pluralistic societies entails not only critically interpreting texts across cultures, media, and genres (Kellner, 2002; New London Group, 1996) but also tailoring language and literacy styles to participate in varied discourse communities (Baker, 2002), thus exemplifying the overlap between digital literacies and Standards-based instruction.

Teacher candidates do not comment on how technology integration enables them to develop critical "consumption" of digital media. This particular skill, being able to use "digital tools to gather, evaluate, and use information," is among the International Society for Technology in Education (ISTE) National Educational Technology Standards (2007), for which all teachers are responsible. Participants do not reference this skill in their reflections on technology use despite having encountered the ISTE Standards in both their FL methodology seminar and Technology in Education course. These teacher candidates seek to be "ac-

---

[3] A predominant focus on the Communication and Cultures goal areas has similarly been reported for in-service teachers (Phillips & Abbott, 2011). This finding is further discussed in the Conclusions and Implications section.

countable practitioners" (Ottesen, 2006, p. 281), which at this point in their professional development entails ensuring student learning of language and culture as discussed above. They do not articulate their role in developing more content-general skills, such as critical thinking or digital literacies, in their future learners. The way in which FL teacher candidates wrestle with technology integration thus aligns closely with the struggles of teacher candidates in other content areas.

As participants consider their future instructional practices, issues such as time, accessibility, and effectiveness surface. Unlike teacher candidates and teachers in previous studies, who found it difficult to bridge the gap between theoretical discussions of technology integration and practice, these future teachers are able to identify ways they would use the presented tools in their own teaching (see comments 3, 9, 10, and 11). This could be due to questionnaire items which asked participants to apply the presented material, which was specific to FL education, to their instructional contexts. Comments from teacher candidates go beyond a simple approval or disapproval of a technological tool, as they clearly wrestle with the tensions of time and accessibility.

(13) Though I absolutely LOVE [place-based mobile gaming], I think that creating it and funding it (buying the technology) may make it too difficult to implement in my classroom. If I could find a way to implement it, I would be quick to do so, but at this point it doesn't seem likely. (Fall 2012)

(14) For the child who does not have internet access at home, how do you differentiate for that? (Panel Discussion Question, Winter 2013; Edmodo, Quia)

(15) [Place-based mobile gaming] sounds like a great idea, but very time consuming and not practical for a high school setting.... I would say 90% or more of my students have a smart phone and/or iPod, so I don't think that is the issue. It would be creating the game, allowing them to go out and do it, and then assuring they are learning from it. (Fall 2012)

(16) [Making a graphic novel] seems like it would be very time-consuming. It also seems like there are a lot of steps involved which would be a concern when teaching 150+ students to use the program. Plus, depending on the school district, you would need to provide in-class time to work on the project. (Winter 2013)

In some instances (such as comment 13), their excitement about the learning potential that accompanies a technology is tempered by the reality that they do not have access to the necessary hardware or software. Although focused on pedagogical implementation of technology, presentations by university faculty inevitably showcased the wealth of resources at higher education institutions. Teacher candidates' comments highlight the discrepancy in available resources between the secondary and postsecondary contexts and reflect a feeling of constraint. An additional limitation of the secondary context as identified by teacher candidates lies in the lack of autonomy of high school students given their younger age and more structured school setting (see comments 15 and 16). Throughout all their comments, these future teachers demonstrate an acute awareness of the role of instructional context and keep student language learning at the forefront.

As seen in the outset of the analysis, teacher candidates seek out efficient and effective ways of teaching and learning and question whether particular tools are the best fit (see comment 1). They see technology as one tool among many for learning, and make "decisions about the use or nonuse of technology ... solely on the basis of what is best for a given teaching situation" (Hubbard, 2008, p. 179). They recognize the call to incorporate technology into instruction yet want to ensure that it truly enhances learning.

> (17) It did seem like [chatrooms] would yield a lot of helpful information and could be great for the students. I just thought it may have been more time consuming; whereas, there are many other activities focused on conversation that could establish the same goals. (Fall 2012)

> (18) I would love to do [graphic novels] as a school project. My only concern is time with this. I would really have to look at what my students would take away from this. (Fall 2012)

> (19) Outside of class it will be hard to formally assess the students and check for complete understanding. Sometimes, students will find their way around the game and be able to "guess" which way to maneuver based on their tech-savvy skills, rather than their language abilities. (Fall 2011; place-based mobile gaming)

As they contemplate how various technologies would play out, teacher candidates acknowledge the technology skills that students already bring to the classroom, likely stemming from their technology use outside of school. Because of this, they feel a need to stay one step ahead of students in order to maintain control over the teaching and learning process and ensure that students are learning the intended content, similar to the findings of Burnett (2011), Ottesen (2006), and Kessler (2010). The teacher candidate in comment 19 views her future students as more "tech-savvy" than she is, even though she could be considered a digital native herself. Whether or not her students are actually more adept in their technology skills, this particular teacher candidate expresses uncertainty about ensuring student learning when using technology. Rather than viewing the skills students bring to the table as advantageous, she perceives them as an obstacle that may interfere with content learning, ignoring how they might facilitate learning or ways to further develop them.

Uncertainty about their own technological skill and how to best incorporate technology into instruction leaves some of the teacher candidates with a desire to play with technology and experience it before using it in their own classrooms; others immediately rule out certain tools because of their perceived lack of ability.

> (20) These are things my students already know how to use but I have never had the opportunity to learn about them. (Winter 2013)

> (21) I would obviously need a tutorial on the [place-based mobile gaming] program. (Winter 2013)

(22) I don't think I have the ability to design such an interactive activity in my classroom. (Fall 2012; place-based mobile gaming)

(23) I liked learning about different technologies because I'm really behind the times on these. (Fall 2011)

The lack of confidence with regard to implementing certain technologies supports the findings from multiple studies on teachers and teacher candidates and their attitudes towards using instructional technologies (e.g., Bosch & Cardinale, 1993; Davis, 1993; Fratianni et al., 1990; Heinich, 1991; Nahachewsky, 2007; Robinson & Mackey, 2006; Topp, 1996; Topp et al., 1995). Comments 19-22 highlight teacher candidates' insecurity regarding their technical competence rather than their uncertainty about effectuating meaningful integration, when in fact research suggests that meaningful integration is the greater challenge (Dutt-Doner et al., 2006; Kessler, 2010).

3.2 The Emerging Professional and Technology Integration

As teacher candidates differentiate their own degree of technological skill from that of their students, we see them begin to separate themselves from their students and their new sense of professional self begins to emerge. The developing professional identity of the teacher candidate, inherent in any field as the "trainee" becomes increasingly knowledgeable and experienced, is echoed in multiple comments above where we see them adopt the perspective of a teacher and express an insider's point of view. Comment 1, repeated here, shows the teacher candidate's identity shifting from that of university student to that of teacher.

(1) I think it can be confusing when [technology] is best and when we need to let it go. I think this is a really important point because our professors really push us to use a lot of technology in everything we do, but sometimes that isn't practical. (Fall 2012)

At this point in their program, opportunities to develop their professional identity tend to be limited to their field placements, yet the dynamics continue to reinforce their role as a preprofessional dependent on the guidance of experienced teachers and university supervisors. In spite of the hierarchical relationship intrinsic to practica, the "prolonged learning-in-practice" (Lave, 1996) that accompanies a field placement offers them a space where they can gain confidence and experience and develop their teacher identity (Kanno & Stuart, 2011). Consequently, teacher candidates are empowered to question instructional principles and engage in critical dialogue with this experience in mind, as highlighted in the following comment:

(24) I think our field experience allows us to ask the best possible questions and see direct application to our classrooms. (Winter 2013)

Participants value the opportunity to interact as professionals, especially within an academic setting, to talk about ideas and tools *as teachers*. In this way, teacher candidates are allowed to further develop their professional identity and began to see themselves not only as teachers but also as members of a larger discourse community, with the ability to take others' ideas and adapt them to their own instructional contexts. The importance of developing teacher candidates' professional identity is well attested in the literature (e.g., Burnett, 2011; Kanno & Stuart, 2011; Ottesen, 2006), and the findings from this study provide further support for this.

> (25) I actually felt like a professional. It was a genuine experience with real experts on the subject matter. [The videoconference] was not a faked activity to teach us about technology. (Winter 2013)
>
> (26) It was really neat to talk to other professionals in the foreign language field about teaching ideas. (Fall 2011)
>
> (27) I felt that it really gave us a chance to speak to people around the country and get an idea of what people in our field are doing (Fall 2011).

The structure of this particular learning experience, modeled after a professional conference panel where speakers present rather than teach, and the fact that the content of the panel presentations is directly related to their concurrent field experiences, allows participants to engage with the content as professionals and further develop their emerging professional identities.

Learning about FL teaching from other professionals, as opposed to from professors in a required course, empowers them not only to critique and adapt the pedagogical activities as seen previously, but also to share them with other teachers in their field placement. Participants recognize that the tools presented were cutting edge and part of the current conversation in the field; the issues discussed are pertinent and relevant to all FL practitioners, not just to them as new members of this community.

> (28) I showed this [open-access digital learning environment] to my [cooperating teacher] and she was so thrilled. She immediately signed up to get information and materials sent to her and she sent out the information about the site to her colleagues in the same department. (Winter 2013)
>
> (29) I appreciated that each speaker was working in the field and using these techniques in the classroom. It was not just theory. (Fall 2012)
>
> (30) I took away a few innovative ideas that weren't in our textbook and I hope to use in the classroom some day. (Fall 2012)
>
> (31) I learned some insight into how teaching a foreign language can still be difficult with professors who research/study teaching techniques, which helped me realize that it's okay to try new projects .... (Fall 2012)

As these teacher candidates consider technology integration, they position themselves as a professional rather than as a student, guided by language learning objectives and general principles of good teaching. In line with previous research on teacher candidates' use of instructional technology (e.g., Bosch & Cardinale, 1993; Davis, 1993; Fratianni et al., 1990), the future teachers in this study identify their own lack of technological skill and the limitations of their instructional contexts as obstacles to effective technology integration. In spite of this, they are enthusiastic about the potential that technology offers for FL teaching and learning.

4. Conclusions and Implications

The findings of this study illustrate the ways in which these FL teacher candidates intend to incorporate technology in their future teaching. However, like teacher candidates in other studies, they voice concern about losing control over learning and the learning environment while implementing technology (Kessler, 2010). As such, in practicing their new identity as an "accountable" (Ottesen, 2006, p. 281) FL teacher, their primary focus is to ensure that language and culture are being learned. This focus on only two of the five goal areas (Communication and Cultures), in line with the current practices of in-service teachers (Phillips & Abbott, 2011), limits the way in which they view technology integration in FL instruction. While they mention aspects of digital literacies, such as collaboration, creativity, and interaction, other elements, such as developing critical thinking and the ability to synthesize information, are noticeably absent from their critiques. Moreover, the aspects they identify are not explored in depth, nor are all of them connected to FL instruction. They are an inadvertent byproduct of technology integration, viewed in isolation rather than guiding planning or instruction.

The position these teacher candidates take toward technology integration resembles that of in-service ESL/EFL teachers as described in Kim (2008). These future FL teachers describe technology in a way that could be characterized as a "supplemental teacher's tool" given that they see it as "a resource, a tool for tutoring, communication, presentation and writing, a motivator, and an optional tool" (p. 255). Technology use for them has not yet been "normalized" (Bax, 2003) because decisions are not based solely on what is best for student learning defined broadly, but more on the availability of tools, teacher comfort with technology, and language and culture objectives. That is not to say that teacher candidates do not consider instructional purpose, but skills not specific to FL learning generally do not figure into their goals for instruction.

Teacher candidates, new to the field, must forge their own instructional practices, basing them on experiences as a FL learner, observations and experiences in field placements, and knowledge and competencies gained from teacher preparation coursework. Previous research suggests that teachers often adopt instructional approaches that resemble the way in which they were taught (e.g., Russell, 1997; Schifter, 1997; Scholz, 1995); for today's FL teacher candidates, their experiences as a learner likely did not reflect "the interdisciplinary purpose of the standards" (Phillips & Abbott, 2011, p. 11) and, by association, did not explore the critical dimensions of digital literacies. As discussed above, pedagogies of digital or multiliteracies have yet to take hold in FL instruction (Hubbard, 2008). Because teacher candidates have not experienced the integration of digital literacies instruction as a learner nor observed it in field placements, their interaction with the construct is limited to their teacher preparation program at the university. Future research might explore how FL teacher candidates' previous experiences and views of digital literacies inform their instructional practices.

Teacher preparation programs commonly require coursework on technology and literacy; yet, in line with findings from other content areas (e.g., Burnett, 2011; Merkley et al., 2001; Nahachewsky, 2007), the findings from this study suggest that today's programs are not adequately raising FL teacher candidates' awareness of the need to address digital literacies, and, consequently, teacher candidates are not prepared to integrate digital literacies into instruction. Within a teacher preparation program, literacy instruction and technology integration often represent two separate competencies, both across courses and within an individual course. Comprehensive treatment of digital literacies may be lacking in FL methods materials in spite of the findings of Arnold (2013) that methods textbooks do "weave" discussion of technology throughout (p. 233). Consequently, while connections can be made between technology and literacy in pedagogy courses, they likely do not play a prominent role or guide instruction, which may influence how teacher candidates plan for digital literacies in their own instruction.

In conclusion, teacher preparation programs can no longer view these two fields as disparate and unrelated but instead must merge them and address them systematically throughout the program. Thus, a broader reconceptualization of language teacher education is essential, one that is based on an expanded view of literacy in FL instruction (Byrnes, 2005; Kern, 2000, 2003), which in today's society inherently includes the digital. As the current study shows, isolated treatment of digital literacies does not appear to raise consciousness enough to impact understanding or instructional practice. While we acknowledge that some competencies take years of teaching experience to develop (Compton, 2009), teacher education programs can plan for a comprehensive approach to raise teacher candidates' awareness of the digital literacy practices, their importance, and ways to address them in instruction. One approach might be to introduce digital literacies in a methods course early on in the term as a way to frame the rest of the experience, followed by regular reflections throughout the term on how digital literacies play out in their instructional practices. In order for today's teacher candidates to break free from the "'old wine in new bottles' syndrome, whereby long-standing school routines have a new technology tacked on here or there, without in any way changing the substance of the practice" (Lankshear & Knobel, 2006, p. 55), teacher preparation institutions must do the same. Future research must then investigate what specific practices at institutions of higher education best prepare FL teacher candidates to adopt digital literacies as a guiding construct in their instruction.

References

Achieve. (2013). *Closing the expectations gap: 2013 annual report on the alignment of state K-12 policies and practice with the demands of college and careers*. Retrieved from http://www.achieve.org/files/2013ClosingtheExpectationsGapReport.pdf

American Council on the Teaching of Foreign Languages. (2002). *ACTFL/NCATE program standards for the preparation of foreign language teachers.* Yonkers, NY: Author.

Arnold, N. (2013). The role of methods textbooks in providing early training for teaching with technology in the language classroom. *Foreign Language Annals, 46,* 230–245.

Baker, J. (2002). Trilingualism. In L. Delpit & J. K. Dowdy (Eds.), *This skin that we speak* (pp. 49–61). New York, NY: The New Press.

Bax, S. (2003). CALL—Past, present, and future. *System, 31,* 13–28.

Bosch, K. A., & Cardinale, L. (1993). Preservice teachers' perceptions of computer use during a field experience. *Journal of Computing in Teacher Education, 10,* 23–27.

Bucci, T. T., Petrosino, A. J., Bell, R., Cherup, S., Cunningham, A., Cohen, S., ... Wetzel, K. (2004). Meeting the ISTE challenge in the field: An overview of the first six distinguished achievement award winning programs. *Journal of Computing in Teacher Education, 21,* 11–21.

Buckingham, D. (2008). Defining digital literacy: What do young people need to know about digital media? In C. Lankshear & M. Knobel (Eds.), *Digital literacies: Concepts, policies and practices* (pp. 73–89). New York, NY: Peter Lang.

Burnett, C. (2011). Pre-service teachers' digital literacy practices: Exploring contingency in identity and digital literacy in and out of educational contexts. *Language and Education, 25,* 433–449.

Byrnes, H. (2005). Literacy as a framework for advanced language acquisition. *ADFL Bulletin, 37*(1), 85–110.

Compton, L. K. L. (2009). Preparing language teachers to teach language online: A look at skills, roles and responsibilities. *Computer Assisted Language Learning, 22,* 73–99.

Conte, C. (1997). *The learning connection: Schools in the information age.* Washington, DC: Benton Foundation.

Cuban, L. (1995). Reality bytes: Those who expect technology to change school will have to wait. *Electronic Learning, 14*(8), 14–15.

Davis, N. (1993). The development of classroom applications of new technology in preservice teacher education: A review of the research. *Journal of Technology and Teacher Education, 1,* 229–249.

De Lange, T., & Skedsmo, G. (2004). *IKT i lærerutdanningen. En studie av digitale læringsvilkår på praktisk pedagogisk utdanning ved universitetet i Oslo* [ICT in teacher education: A study of digital learning possibilities at the University of Oslo]. Unpublished manuscript, Department of Teacher Education and School Development, Oslo University, Norway.

Driskell, T. (1999). *The design and development of Helper, a constructivist lesson plan web resource to model technology integration for teachers* (Unpublished doctoral dissertation). University of Houston-Central Campus, Houston, Texas.

Dutt-Doner, K., Allen, S. M., & Corcoran, D. (2006). Transforming student learning by preparing the next generation of teachers for type II technology integration. *Computers in the Schools, 22,* 63–75.

Fernández Dobao, A. (2012). Collaborative dialogue in learner-learner and learner-native speaker interaction. *Applied Linguistics, 33,* 229–256.

Fratianni, J., Decker, R., & Korver-Baum, B. (1990). Technology: Are future teachers being prepared for the 21st century? *Journal of Computing in Teacher Education, 6,* 15–23.

Fulton, K. (1998). Learning in the digital age: Insights into the issues. The skills students need for technological fluency. *THE Journal, 25*(7), 60–63.

Gee, J. P. (2007). *Good video games + good learning: Collected essays on video games.* New York, NY: Peter Lang.

González, N., Moll, L., & Amanti, C. (2005). *Funds of knowledge: Theorizing practices in households, communities, and classrooms.* Mahwah, NJ: Erlbaum.

Graham, L. (2008). Teachers are digikids too: The digital histories and digital lives of young teachers in English primary schools. *Literacy, 42,* 10–18.

Heinich, R. (1991). Restructuring, technology, and instructional productivity. In G. Anglin (Ed.), *Instructional technology: Past, present, and future* (pp. 236–243). Englewood, CO: Libraries Unlimited.

Hubbard, P. (2008). CALL and the future of language teacher education. *CALICO Journal, 25,* 175–188.

International Society for Technology in Education. (2007). *ISTE standards-students.* Retrieved from http://www.iste.org/docs/pdfs/20-14_ISTE_Standards-S_PDF.pdf

Jensen, C. (2003). Pluto-5. PPU og det eksemplariske prinsipp. Evaluering av Pluto-prosjektet som bidrag til PPU-studenters utvikling av lærerprofesjonalitet [Pluto-5. PPU and the principle of exemplar. Evaluation of the Pluto project as contribution to PPU students' development of teacher professionalism]. Unpublished manuscript, Department of Teacher Education and School Development, Oslo University, Norway.

Kanno, Y., & Stuart, C. (2011). Learning to become a second language teacher: Identities-in-practice. *Modern Language Journal, 95,* 236–252.

Kassen, M., & Lavine, R. (2007). Developing advanced level foreign language learners with technology. In M. Kassen, R. Lavine, K. Murphy-Judy, & M. Peters (Eds.), *Preparing and developing technology-proficient L2 teachers* (pp. 233–262). San Marcos, TX: CALICO.

Kellner, D. (2002). Technological revolution, multiple literacies, and the restructuring of education. In L. Snyder (Ed.), *Silicon literacies: Communication, innovation, and education in the electronic age* (pp. 154–169). New York, NY: Routledge.

Kern, R. (2000). *Literacy and language teaching.* Oxford, England: Oxford University Press.

Kern, R. (2003). Literacy and advanced foreign language learning: Rethinking the curriculum. In H. Byrnes & H. H. Maxim (Eds.), *Advanced foreign language learning: A challenge to college programs* (pp. 2–18). Boston, MA: Thomson Heinle.

Kessler, G. (2010). When they talk about CALL: Discourse in a required CALL course. *CALICO Journal, 27,* 376–392.

Kim, H. K. (2008). Beyond motivation: ESL/EFL teachers' perceptions of the role of computers. *CALICO Journal, 25,* 241–259.

Lankshear, C., & Knobel, M. (2006). *New literacies: Everyday practices and classroom learning* (2nd ed.). New York, NY: Open University Press.

Lave, J. (1996). Teaching, as learning, in practice. *Mind, Culture, and Activity, 3,* 149–164.

Lee, L. (2004). Learners' perspectives on networked collaborative interaction with native speakers of Spanish in the US. *Language Learning & Technology, 8,* 83–100.

Livingstone, S., van Couvering, E., & Thumin, N. (2005). *Adult media literacy: A review of the research literature.* London, England: Ofcom.

Martin, A. (2005). DigEuLit—a European framework for digital literacy: A progress report. *Journal of eLiteracy, 2,* 130–136.

Mayo, N. B., Kajs, L. T., & Tangum, J. (2005). Longitudinal study of technology training to prepare future teachers. *Educational Research Quarterly, 29,* 3–15.

Merkley, D. J., Schmidt, D. A., & Allen, G. (2001). Addressing the English Language Arts technology standard in a secondary reading methodology course. *Journal of Adolescent and Adult Literacy, 45,* 220–231.

Nahachewsky, J. (2007). At the edge of reason: Teaching language and literacy in a digital age. *E-Learning, 4,* 355–366.

National Governors Association Center for Best Practices, Council of Chief State School Officers. (2010). *Common Core State Standards for English Language Arts and Literacy.* Washington, DC: Author. Available from http://www.corestandards.org/wp-content/uploads/ELA_Standards.pdf

National Standards in Foreign Language Education Project. (2006). *Standards for foreign language learning in the 21st century.* Lawrence, KS: Allen Press.

New London Group. (1996). A pedagogy of multiliteracies: Designing social futures. *Harvard Educational Review, 66,* 60–93.

Omaggio Hadley, A. (2000). *Teaching language in context* (3rd ed.). Boston, MA: Heinle.

Ottesen, E. (2006). Learning to teach with technology: Authoring practised identities. *Technology, Pedagogy, and Education, 15,* 275–290.

Ottesen, E. (2007). Teachers "in the making": Building accounts of teaching. *Teaching and Teacher Education, 23,* 617-623.

Phillips, J. K., & Abbott, M. (2011, October). *A decade of foreign language standards: Impact, influence, and future directions.* Report of Grant Project #P017A080037, Title VII, International Research Studies, US Department of Education to the American Council on the Teaching of Foreign Languages. Retrieved from http://www.actfl.org/sites/default/files/pdfs/public/national-standards-2011.pdf

Pool, C. (1997). A conversation with Paul Gilster. *Educational Leadership, 55,* 6–11.

Prensky, M. (2001). Digital natives, digital immigrants. Part 1. *On the Horizon, 9*(5), 1–6.

Purcell-Gates, V. (2002). Multiple literacies. In B. J. Guzzetti (Ed.), *Literacy in America: An encyclopedia of history, theory, and practice* (pp. 376–380). Santa Barbara, CA: ABC-CLIO.

Reinhardt, J., & Thorne, S. L. (2011). Beyond comparisons: Frameworks for developing digital L2 literacies. In N. Arnold & L. Ducate (Eds.), *Present and future promises of CALL: From theory and research to new directions in language teaching* (pp. 257–280). San Marcos, TX: CALICO.

Robinson, M., & Mackey, M. (2006). Assets in the classroom: Comfort and competence with media among teachers present and future. In J. March & E. Millard (Eds.), *Popular literacies, childhood and schooling* (pp. 200–220). London, England: Routledge.

Russell, T. (1997). Teaching teachers: How I teach is the message. In J. Loughran & T. Russell (Eds.), *Teaching about teaching: Purpose, passion and pedagogy in teacher education* (pp. 32–47) London, England: Falmer.

Sawhill, B. (2008). The changing role of the language teacher/technologist. Connected learning, meaningful collaborations, and reciprocal apprenticeships in the foreign language curriculum. *IALLT Journal of Language Learning Technologies, 40,* 1–17.

Schifter, D. (1997). *Learning mathematics for teaching: Lessons in/from the domain of fractions.* Newton, MA: Educational Development Center, Inc. (ERIC Document Reproduction Service No. ED412122)

Scholz, J. M. (1995, April). *Professional development for mid-level mathematics*. Paper presented at the annual meeting of the American Educational Research Association, San Francisco, CA. (ERIC Document Reproduction Service No. ED 395 820).

Sheridan-Thomas, H. K. (2007). Making sense of multiple literacies: Exploring pre-service content area teachers' understandings and applications. *Reading Research and Instruction, 46,* 121–150.

Shotter, J. (1984). *Social accountability and selfhood*. Oxford, England: Blackwell.

Shrum, J. L., & Glisan, E. W. (2010). *Teacher's handbook: Contextualized language instruction*. (4th ed.). Boston, MA: Heinle.

Thorne, S. L., & Reinhardt, J. (2008). "Bridging activities," new media literacies and advanced foreign language proficiency. *CALICO Journal, 25,* 558–572.

Topp, N. (1996). Preparation to use technology in the classroom: Opinions by recent graduates. *Journal of Technology and Teacher Education, 12,* 24–27.

Topp, N.W., Mortensen, R., & Grandgenett, N. (1995). Building a technology-using facility to facilitate technology-using teachers. *Journal of Computer in Teacher Education, 11*(3), 11–14.

van Compernolle, R. A., & Williams, L. (2009). (Re)situating the role(s) of new technologies in world language teaching and learning. In R. Oxford & J. Oxford (Eds.), *Second language teaching learning in the Net Generation* (pp. 9–22). Honolulu: University of Hawai'i, National Foreign Language Resource Center.

Appendix

Survey Questions

1. Of all the different technologies explored in class (both in class and during the panel),
   a. Which do you think you will be most likely to use in your teaching? Why? How do you see yourself using this technology?
   b. Which do you think you will be least likely to use in your teaching? Why?
2. What were your reactions to the panel discussion/presentation as part of the class? What did you like about it? What did you learn from it? What would you change about it?
3. What were your reactions to using video conferencing in a classroom situation?

Index

activity theory 180–182, 186
affordances 5, 17, 25, 34, 88, 90–91, 107, 123–124, 126–128, 132, 136, 163–164, 201–203, 205, 208, 214, 220, 222, 228
agency 2, 16, 19, 87, 90, 104, 106–108, 122–123, 146, 161–162, 169, 171, 182, 228, 259
American Council on the Teaching of Foreign Languages (ACTFL) 122, 270
collaborative 3, 9, 14–15, 17, 21–23, 25, 106, 119, 121, 126, 129, 131, 160, 181, 183–184, 187, 190, 202–206, 208, 212–213, 216, 220–222, 229, 246, 267, 271
collaboration 4, 12, 18, 22, 105, 119, 121, 125, 131, 133–134, 180, 215, 222, 229, 272, 274, 279
Common Core State Standards Initiative 266
communicative competence 3, 9–10, 12–13, 15, 19–21, 23, 25, 30, 122
computer literacy 33, 50, 71, 104, 143
computer–mediated communication (CMC) 32, 36, 204
computer–mediated enterprise 221
computer–mediated semiotic activity 205
computer–mediated Spanish writing environment 221
computer–mediated FL writing environment 184, 187
digital age 44, 69, 72, 203–204, 267
digital divide 4, 80, 119, 123, 135
digital era 41, 143
digital game 160
digital gaming 159–160, 162–163, 167–168, 170–172
digital immigrant(s) 3, 30, 69, 72–75, 79

digital native(s) 3, 21, 29–32, 35–36, 44–48, 52, 57–59, 65, 69, 72–75, 79, 123, 157, 222, 228, 242, 265, 276
digital nativeness 32, 45, 48, 49, 57
digital world(s) 21, 46, 50, 266–267
discourse 4, 9, 10, 14, 17, 19, 25, 30, 91, 94, 108, 126–128, 130, 142, 161–162, 164–165, 167–169, 171, 197, 205, 221, 227–228, 270, 274, 278
ecological 164, 202
first language (L1) 5, 37, 42, 120, 124, 127–128, 141, 146, 154, 162, 183, 203, 212, 214, 216, 227, 229–230, 234, 240–241
framework 3, 9–10, 19, 94, 119–120, 122, 125, 135, 143, 156, 160–166, 181, 186, 203, 222
game literacies 4, 159, 160–165, 167
Gee, J. 2, 90, 91, 160–161, 163–164, 168, 214, 266
globalization 12, 13, 20, 24, 155, 188
glocal 12–13, 18, 20–21, 24
hardware 52, 57–58, 275
heritage speaker(s) 180, 183–184
hypertext 30, 73, 228
identity(-ies) 3, 4, 14, 124–126, 129, 131, 133–134, 141–157, 161, 168, 170, 268, 277–279
information and communication technology (ICT) 2, 30, 69–72, 75–81
information literacy 143, 228, 246
International Society for Technology in Education (ISTE) 274
Kern, R. 1, 2, 91, 159, 160, 163, 166, 269, 280
Knobel, M. 2, 30, 121, 143, 161, 246, 248–249, 280
L1 (see *first language*)
Lankshear, C. 2, 16, 30, 121, 143, 161, 246, 248–249, 280
media literacy 119–121, 161–162, 228, 235

microblogging 14, 35, 76, 209
microjournal 21
motivation(s) 4, 33, 87–88, 90, 94–95, 103–104, 106–108, 123, 150, 183, 187, 191–192
multilingual 4, 19, 21, 24, 123, 141–143, 145–148, 150, 154–157, 163, 222, 229, 240
multiliteracies 2, 91, 120, 122, 161, 164, 166–167, 217, 228, 267, 269, 279
multimedia 3–4, 9, 14, 17–19, 21, 24–25, 53, 66, 70–71, 92, 127, 132, 135, 143, 180, 197, 207, 211, 213, 222, 226, 246
multimodal 1–4, 9–10, 12, 14–17, 19, 30, 88, 92, 106–107, 119–123, 125, 127–128, 130, 132–136, 141–152, 154–157, 171, 183, 195, 228, 235
National Council for Accreditation of Teacher Education (NCATE) 270
National Council of Teachers of English (NCTE) 161
native language 21, 30, 32, 36, 48, 72, 184, 227, 230, 272–273
native speaker(s) 21, 30, 32, 36, 48, 72, 184, 227, 230, 272–273
new literacies 3, 30, 141, 159–160, 184, 222, 267
New London Group 2, 91, 120, 122, 159, 160, 228, 274
Prensky, M. 3, 21, 30–31, 69, 72–75, 79–80, 121, 265
reading 1, 4–5, 9–12, 15–17, 30, 50, 52, 72–73, 88, 90–94, 100–102, 114, 116, 125, 143, 150, 162, 166–167, 186, 192, 202–222, 228, 235, 246, 274
scaffolding 5, 101, 108, 120, 127, 134
shuttling 130–131
sociocultural 1, 2, 87, 90–91, 95, 107, 121–122, 166, 181–182, 212, 242, 269

software 15, 38, 50, 54, 57–58, 66, 70, 81, 97, 143, 167, 193, 196, 206, 211, 218–219, 228–229, 231, 235, 275
standards (ACTFL) 122, 270, 274, 279
standards (Common Core) 266
standards (ISTE) 270
standards (NCATE) 274
technological literacy 146–147, 151–152, 154, 196, 246
technology literacy 193
Vygotsky, L. 90, 94, 162, 212
web 2.0 9, 14, 32, 80, 87–90, 105–108, 129, 131
writing 1, 2, 4, 5, 9–12, 15–17, 22, 30, 88–92, 94, 98–100, 102–106, 113, 116, 143–144, 150, 156, 164–168, 170, 179–181, 183–187, 189–197, 203–204, 228–229, 234, 245–248, 250, 252–259, 274, 279